'This book by long standing researchers in the field provides a situated and generous view of histories of education and the environment globally. Highlighting the need for cultural shifts and the role of education in that, the book examines whether and how both policy and research have made a difference over the past 50 years. Essential reading for those wanting to understand perspectives on the past and possible future contributions of education and the environment'.

Dr Marcia McKenzie, *Professor in Global Studies and International Education, University of Melbourne and Director, Monitoring and Evaluating Climate Communication, and Education Project (MECCE)*

'We live in an era of poly-crises and are in need of transitions toward a more sustainable world. There are no blueprints for this immense endeavour; instead we must explore and learn our way out of unsustainable living. Education for Sustainable Development (ESD) can help us with the knowledge, tools and means for that.

This book provides us with the history, theories, practices and examples from past and present in how we can apply 'learning' as a strategy for a better future.

We stand upon the shoulders of great educators and organisations who 'walked the talk' by bringing the head, heart and hands into meaningful change towards the future'.

Roel van Raaij, *Steering Committee of the national Dutch ESD Program DuurzaamDoor*

'Elevated knowledge and awareness as to the scale and urgency of ecological crisis that is upon is only one aspect of what is needed in changing our impact. Understanding what has been achieved and what more needs to be done is an essential starting point and this book offers a brilliant and timely summary of exactly that!'

Dr Tony Juniper CBE, *award winning environmentalist, writer and Chair of Natural England*

'Without nature, there is no future; without education, there is no understanding. This book poignantly explores the past, helps us meet the future, reveals success stories in safeguarding the environment and elegantly demonstrates that environmental education is the incubator that will solve complex issues and safeguard this planet. A must read!'

Jan-Gustav Strandenaes, *Senior Adviser on Governance, the Environment and Sustainable Development, Stakeholder Forum for a Sustainable Future*

Education and Learning for Sustainable Futures

Responding to growing interest in the sustainable development goals (SDGs) and global concern over climate change, this volume provides an analysis of how our understanding of the relationship between environment and education has evolved during the past 50 years.

Spanning from the 1972 United Nations Conference on the Human Environment through to the present day, chapters examine whether our approach to education about environmental sustainability is enacting effective change. Examining the evolution of educational approaches to environmental learning, contexts, and themes, this book moves through the decades, from the 1970s until the 2020s, tracking the impact of historical events and shifting sustainability discourses within education. Through historical, research-based analyses, this book recognises patterns, trends, and countertrends that help critically (re)assess the potential of education in creating a world that is more sustainable than current scientific predictions estimate.

Proposing a set of key considerations for the future of environmental education, this accessible book will be of value to scholars, researchers, policymakers, and practitioners working within sustainability education, environmental research and policy, and teacher education more broadly.

Thomas Macintyre is a researcher in the field of education and sustainability, specialising in transformative and participatory learning.

Daniella Tilbury is an Honorary Fellow of St Catharine's College, University of Cambridge, a European Commission advisor on learning for the Green Transition and the UK government's focal point at the UN Economic Commission for Europe on matters of education and environment.

Arjen Wals is a professor of Transformative Learning for Socio-Ecological Sustainability at Wageningen University where he also holds the UNESCO Chair of Social Learning and Sustainable Development.

Routledge Research in Education, Society and the Anthropocene

This series offers a global platform to engage scholars in continuous academic debate on key challenges and the latest thinking on education and society and the role of education in the Anthropocene. It provides a forum for established and emerging scholars to discuss the latest debates, issues, research and theory across the field of education research relating to society and the Anthropocene.

Rethinking Education in Light of Global Challenges
Scandinavian Perspectives on Culture, Society and the Anthropocene
Edited by Karen Bjerg Petersen, Kerstin von Brömssen,
Gro Hellesdatter Jacobsen, Jesper Garsdal, Michael Paulsen
and Oleg Koefoed

Ecosophy and Educational Research for the Anthropocene
Rethinking Research through Relational Psychoanalytic Approaches
Alysha J. Farrell

Justice and Equity in Climate Change Education
Exploring Social and Ethical Dimensions of Environmental Education
Edited by Elizabeth M. Walsh

Education as Human Knowledge in the Anthropocene
An Anthropological Perspective
Christoph Wulf

Engaging with Environmental Education through the Language Arts
Interdisciplinary and Creative Approaches to Fostering Ecoliteracy
Edited by Nicholas McGuinn and Amanda Naylor

Education and Learning for Sustainable Futures
50 Years of Learning for Environment and Change
Thomas Macintyre, Daniella Tilbury and Arjen Wals

Education and Learning for Sustainable Futures

50 Years of Learning for Environment and Change

Thomas Macintyre, Daniella Tilbury, and Arjen Wals

Routledge
Taylor & Francis Group
LONDON AND NEW YORK

First published 2025
by Routledge
4 Park Square, Milton Park, Abingdon, Oxon OX14 4RN

and by Routledge
605 Third Avenue, New York, NY 10158

Routledge is an imprint of the Taylor & Francis Group, an informa business

© 2025 Thomas Macintyre, Daniella Tilbury and Arjen Wals

The right of Thomas Macintyre, Daniella Tilbury, and Arjen Wals to be
identified as authors of this work has been asserted in accordance with sections
77 and 78 of the Copyright, Designs and Patents Act 1988.

Open Access conversion made possible by DuurzaamDoor Onderwijs funded
by the government of The Netherlands.

Trademark notice: Product or corporate names may be trademarks or
registered trademarks, and are used only for identification and explanation
without intent to infringe.

British Library Cataloguing-in-Publication Data
A catalogue record for this book is available from the British Library

Library of Congress Cataloging-in-Publication Data
Names: Macintyre, Thomas, author. | Tilbury, Daniella, author. |
Wals, Arjen E. J., author.
Title: Education and learning for sustainable futures : 50 years
of learning for environment and change / Thomas Macintyre,
Daniella Tilbury, and Arjen Wals.
Description: Abingdon, Oxon ; New York, NY : Routledge, 2025. |
Series: Routledge research in education, society and the anthropocene |
Includes bibliographical references and index.
Identifiers: LCCN 2024040934 (print) | LCCN 2024040935 (ebook) |
ISBN 9781032727912 (hardback) | ISBN 9781032739632 (paperback) |
ISBN 9781003467007 (ebook)
Subjects: LCSH: Environmental education. | Sustainability--Study and
teaching. | Sustainable Development Goals (Project)
Classification: LCC GE70 .M33 2025 (print) | LCC GE70 (ebook) |
DDC 304.2071--dc23/eng/20240925
LC record available at https://lccn.loc.gov/2024040934
LC ebook record available at https://lccn.loc.gov/2024040935

ISBN: 978-1-032-72791-2 (hbk)
ISBN: 978-1-032-73963-2 (pbk)
ISBN: 978-1-003-46700-7 (ebk)

DOI: 10.4324/9781003467007

Typeset in Times New Roman
by KnowledgeWorks Global Ltd.

To those educators – both human and non-human – who lead us towards a better tomorrow, and to our children Mateo, Alexa, Kendra, and Brian, that they may experience a more intergenerational, just and sustainable future.

Contents

Figures

About the Authors

Dr. Thomas Macintyre is a researcher in the field of education and sustainability, specialising in transformative and participatory learning. Following his doctoral research into community-based learning in Colombia, South America, Thomas has worked as a UNESCO research fellow and consultant on UNESCO projects around themes of sustainability, climate change, and education. Thomas has published widely in the field of education and sustainability, with a particular focus on exploring 'transgressive' forms of learning which critically address (un)sustainable norms and worldviews, while proposing alternative visions and practices in education. At his current position at the University of Stavanger in Norway, Thomas is working on citizen involvement in urban development as part of the EU Horizon project NEBSTAR.

Professor Daniella Tilbury is an academic leader, change-maker, and educator in sustainability, credited with having developed the initial frameworks for learning in this area. She is currently an Honorary Fellow of St Catharine's College, University of Cambridge, a European Commission adviser on learning for the Green Transition, and the UK government's focal point at the UN Economic Commission for Europe on matters of education and environment. Daniella chaired the UNESCO Global Monitoring Expert Group in ESD 2004–2015. She served as a member of the Board of WWF Australia that established EarthHour in 2007 and has been actively involved in international initiatives that inspire transitions towards climate-positive futures. She has served as a Research Chair, Dean, Vice-Chancellor, and Governor at Universities in Australia, Hong Kong, Gibraltar and the UK and holds an honorary doctorate from the University of Girona.

Arjen Wals is Professor of Transformative Learning for Socio-Ecological Sustainability at Wageningen University where he also holds the UNESCO Chair of Social Learning and Sustainable Development. Furthermore, he is a Guest Professor at the Norwegian University for the Life Sciences (NMBU). He holds an Honorary Doctorate from Gothenburg

University in Sweden. Arjen is co-founder of Caretakers of the Environment International (CEI). His work focuses on enabling, supporting, and assessing ecologies of learning that foster sustainable living by inviting more relational, ethical, and critical ways of knowing and being. Much of his work focuses on the development of Whole School and Whole University Approaches to sustainability and the decolonisation of education. He writes a regular blog that signals developments in the emerging field of sustainability education: www.transformativelearning.nl.

Acknowledgements

It is with deep gratitude that we acknowledge those who have been instrumental in shaping our perspectives on the environment, education, and sustainable development. Their influence on our pursuit for just and sustainable futures has been an ever-present guiding force, shaping not only our thoughts but also our actions.

First, we extend our sincerest gratitude to the trailblazers in the field of education and the environment. It would be impossible to name them all, though you will read about some of them in this book. That said, this publication has been made possible thanks to the valuable work and assistance from colleagues at the *Stakeholder Forum for a Sustainable Future*, the *Norwegian Forum for Development and Environment*, and the Civil Society Unit at The *United Nations Environment Programme* (UNEP). The *Swedish Government, Ministry of Environment*, provided financial support that enabled us to research the historical roots of Environmental Education and sustainability learning and for this funding, we are most grateful.

We extend our sincere thanks to Jan-Gustav who was the initiator and chief editor of the 'The People's Environment Narrative, the PEN, 50 years with UNEP and Civil Society'. We were invited to contribute a chapter to the PEN, which was presented at the Stockholm+50 conference in June 2022, and documents 50 years of global work to protect and safeguard the environment. The PEN covers more than 900 pages and involves 100 activists, scholars, teachers, practitioners, and others from all continents who have presented their insights in nearly 150 articles. Additionally, thousands of stakeholders contributed to its final content through global webinars which reviewed the initial findings of this work. Seen largely from the point of view of civil society, the PEN is a compendium of articles with encyclopaedic ambitions. The chapter we contributed to the PEN forms the foundation of this book.

Second, we give heartfelt appreciation to our families, whose unwavering support and encouragement have been the bedrock upon which this endeavour rests. Their sacrifices and understanding during late nights of research and writing have not gone unnoticed. This book stands as a testament to the love and strength that family provides, grounding us in our mission to contribute positively to the world. Third, we wish to acknowledge the importance of the non-human in everything we do. The text serves as a nod to the intricate web of life that encompasses not only our fellow human beings but also the flora and fauna that share this planet, the rivers, mountains, and seas, as well as everything else which escapes our understanding. Through this text, we emphasise the significance of respecting and preserving the delicate balance of our ecosystems. As we delve into discussions on the evolving nexus of education and the environment, we recognise the immense importance of education and learning to individuals, communities, and societies to prioritise the health of our planet. Education is not merely a tool for personal advancement; it is a means to empower individuals and communities to carve alternative futures and regenerate our relationship with the environment and life, undoing the exploitative relationships that exist today.

Last, we are grateful to the Duurzaam-Door Onderwijs Program, supported by the Dutch Ministry of Agriculture, Nature and Food Quality for providing the funding for making this publication Open Access and thereby accessible for free across the globe. We take this opportunity to thank the senior policy-advisor of the same Ministry, Roel van Raaij who has spent his entire career advancing environmental and sustainability education in The Netherlands and internationally. Roel has been a colleague, a friend, and a key influencer in the field.

Preface

In a world of climate vulnerability and ecoanxiety, where the delicate balance of ecosystems teeters on the brink, a fundamental question arises: How can we learn to live more sustainably as a global community? While there are many opinions on how this can be done, a common view is that this will involve changing the way we relate to the environment and others. This book explores in detail the last 50 years of education and learning for the environment, helping us make sense of our engagement with the planet. We will be looking back and exploring lived experiences, research findings, and lessons learnt over the last five decades, before turning our gaze forwards to how we can learn towards more sustainable futures.

This publication is the result of a collaborative effort involving three authors with extensive expertise and experience in the fields of education, environment, and sustainability. Writing this book was a multi-step process of reading, classifying, and discussing diverse written material which we considered has contributed to the education and environment nexus. An important contribution has also been made through an online consultation held on May 3, 2022, that was organised by Stakeholder Forum for a Sustainable Future and ForUM Norway on assignment from the Civil Society Unit of UNEP. Participants from numerous countries around the globe discussed the status and position of education and learning in relation to the environment. This dialogue helped unpack the dialectical relationship that has existed between environmental concerns and learning, helping to inform the direction of this book.

We are aware that the process of deciding what to include and exclude in this text is influenced by our own preconceptions, biases, and life experiences. To address this limitation, we have engaged in a reflexive discourse, critiquing amongst ourselves the material we have engaged with, and receiving feedback from colleagues in the field. We have sought to question how our own experiences may create blind spots and limit our engagement with this material, while also recognising that this text marks the beginning of a process. Colleagues and peers in the field of environmental and sustainability education might find the subtitle of this book – 50 Years of Learning for Environment and Change – overly ambitious, and perhaps it is. We want to declare upfront that this book is therefore not a rigorous systematic review of all available literature in multiple languages. Rather, it is a well-informed inquiry, based on some key literature (available mostly in English) and our combined experiences in the field; some of which go back 40 years and spans numerous continents.

In advance, we wish to acknowledge that our lens is mainly a Western or Northern one, rooted in our own histories that in many ways, often subliminally, are intertwined with colonial ways of thinking and acting. Thomas Macintyre was born in New Zealand and was initially shaped by the British schooling system, but has been living in Colombia, South America, for the last 10 years, and has been influenced by alternative worldviews that challenge dominant Western paradigms. Daniella Tilbury is Gibraltarian and has therefore been brought up with Mediterranean ways of seeing the world, but has been influenced by British perspectives as well as experiences of living in Australia and Asia for over a decade. Her formative years in the field were in the 1990s when she read for a Ph.D. on environmental education for sustainability at the University of Cambridge. Daniella has travelled extensively across Africa, Asia and Latin America to review Environmental Education programmes for the World Wide Fund for Nature (WWF). She now works with international policy frameworks. Arjen Wals was born and raised in The Netherlands but lived for many years in the United States while doing his Ph.D. work in inner-city schools in Detroit, Michigan with Bill Stapp, a founding father of Environmental Education, as his mentor. He also has been working in Sweden and Norway on Environmental and Sustainability Education and has worked on a wide range of development projects in Africa, Asia, and Latin America. So even though we have been exposed to non-Western ways of seeing the world, we cannot claim to properly represent other viewpoints, such as decolonial or indigenous perspectives. We have, however, made an effort to include non-Western currents in education and other ways of thinking. While this by no means implies that we have overcome this bias or shortcomings, we do hope that readers will be aware of our efforts and limitations and that it will open up conversations about other ways of framing the issues raised.

We invite you and others interested in understanding the roots of environmental learning and education to assist us in mapping the trajectory that has shaped our current understanding in the field. We believe this work can help us learn from experience, question educational practices, and see more clearly the intentions and future opportunities for learning and education in this area.

1 Introducing 50 Years of Education and Learning for the Environment and Sustainable Futures

Prelude

Human dependency on the environment and entanglement with nature has been historically recorded through the folklore of numerous cultures, ethnic groups, and tribes (Botzler & Armstrong, 1993; Naess, 1990; Weston, 1999). Evidence suggests that already 65,000 years ago, Neanderthals were representing the world around them through depicting animals, human hands, and clubs in cave paintings (Hoffmann et al., 2018). Over the following millennia, a great diversity of environmental knowledge systems and beliefs have developed in cultures around the world (Selin, 2003). The worldviews of these groups often embrace holistic connections between humans and nature, emphasising harmony, reciprocity, and sustainable coexistence with the environment, acknowledging the interconnectedness of all living beings (Botha et al., 2021; Muller et al., 2019).

Religious influences have also shaped human engagement with and perceptions of nature. Mediaeval cosmologies were informed by ideas that God had designed the natural world and fostered harmonious man-environment relationships (Pepper, 1984). Lovejoy (1974) and other writers observed how Christian teachings then shifted views as teaching rested on the assumptions that nature was created 'for man's sake'. White (1967) argues that this anthropocentric sentiment underpinned the Christian-Judaeo doctrine from the Middle Ages and influenced the modern development paradigm; we have since come to understand that this worldview thrives through the exploitation of nature and natural resources.

Other Western influences can be traced back to the early 18th century, with influential figures such as Jean-Jacques Rousseau writing about the importance of education focussed on the environment, providing the foundation of nature study. In the 1920 and 1930s, conservation issues began to emerge from the great depression in the United States, which saw a shift from the study of nature and natural history, towards the study of phenomena that affected both nature and agriculture, such as the destructive sandstorms of the 1930s of the American prairies, also referred to as the dust bowl (McLeman et al., 2014).

DOI: 10.4324/9781003467007-1

The 1950s and 1960s brought increasingly strong signals that water, soil, and air quality were decreasing, and that this was affecting human health. Rachel Carlson's book the *Silent Spring* (1962), on the detrimental effects of pesticide use, was a wake-up call on increasing environmental degradation (Carson, 1962). Likewise, the book *Limits to Growth*, by the Club of Rome, provided stark, scientific warnings that the Earth's resources would not be able to support the current, exponential rates of economic growth and population (Meadows et al., 1972). With global environmental pressures growing, organisations such as the World Wide Fund for Nature and Greenpeace began sounding the alarm bells (Hicks, 2012). In the complex, political context of the Cold War, Schumacher's influential yet simply communicated text 'Small is Beautiful' captured the socio-economic underpinnings of the ecological crisis and promoted a practical approach to addressing the issues. The text showed the interplay that exists between environmental, social, and economic concerns, locating debates outside of ecological and technical discussions which had dominated to date. It also referred to education 'as the greatest resource' (Schumacher, 1973, p. 64). It was within this context that Sweden proposed what would become the United Nations Conference on the Human Environment, in Stockholm, and which took place between June 5 and 16, 1972.

The 1972 UN Conference on the Human Environment

The 1972 Stockholm conference provides the starting point for this book. As the first of the environmental 'mega conferences', Stockholm 1972 was fundamental to providing a reflection on the overall trajectory of human development and its relationship to the environment as a whole (Seyfang, 2003). It was the first global meeting to recognise the interconnections between development, poverty, and the environment and saw a large presence and influence of non-state actors, including non-governmental organisations (NGOs) and scholars. It was also ground-breaking in that it sought global policy consensus on issues related to the environment (Najam & Cleveland, 2005). Records document, however, that it was a contentious meeting: most Soviet bloc countries boycotted the meeting due to the exclusion of then East Germany. There was also strong scepticism from developing countries who were apprehensive of the global North's environmental focus and were concerned about how this would override their human development priorities (Najam & Cleveland, 2005).

Despite these obstacles, the 1972 Stockholm conference was successful in developing a global environmental discourse. Participants adopted a series of principles for sound management of the environment, including the Stockholm Declaration and Action Plan for the Human Environment (Handl, 2012). The ideas in these documents have been carried forth to subsequent summits. Another important result was the establishment of the annual 'World Environment Day', which is now observed in most countries each June and has a strong focus on environmental awareness, learning, and engagement. The 1972 Stockholm Declaration, which contained 26 principles, placed environmental issues at the forefront of international concerns. It marked the start of a dialogue between industrialised and developing countries on the link between economic growth, the pollution of the air, water, and oceans and the well-being of people around the world, highlighting the finite nature of Earth's resources and the necessity for humanity to safeguard them. The major institutional legacy was the creation of the United Nations Environmental Programme (UNEP). Stockholm also cemented the importance of the environment on the international agenda, and through its principle 19, identified education as an environmental strategy, laying the foundation for the Environmental Education movement (see Figure 1.1).

The 1972 Stockholm conference marked a watershed moment for education, providing a platform and catalysing global awareness of environmental

'Education in environmental matters, for the younger generation as well as adults, giving due consideration to the underprivileged, is essential in order to broaden the basis for an enlightened opinion and responsible conduct by individuals, enterprises, and communities in protecting and improving the environment in its full human dimension. It is also essential that mass media of communications avoid contributing to the deterioration of the environment, but, on the contrary, disseminate information of an educational nature on the need to protect and improve the environment in order to enable man to develop in every respect'.

(Principle 19)

Figure 1.1 Principle 19 of the 1972 Stockholm Declaration

Image source: United Nations

challenges. It also emphasised the integral role of education in addressing environmental concerns, establishing environmental governance structures and principles, as well as positioning education as an international strategy requiring collaborative action for the protection of the planet. Today, as we review Stockholm's legacy, we must also acknowledge critiques that few *concrete* changes occurred after Stockholm 1972, and other major conferences (including Rio 1992), despite the widespread support for the pledges, principles, and growing support for the environment. As noted by Clarke and Timberlake (1982), UNEP met in Nairobi for a 'Stockholm+10' conference and similarly concluded that small steps had been taken in reaching the declaration's goals.

Stockholm+50: Where We Are Now

Half a century later, the 2022 Stockholm+50 conference (see Figure 1.2) provided a critical opportunity to ask the question: W*hat have we learnt in half a century about the role of education and learning in relation to the environment and sustainability?* On the one hand, we can celebrate 50 years of global environmental legislation and governance that, for example, has led to the banning of CFCs that has helped mend the hole in the Ozone layer (Leinfelder, 2013) and has improved the quality of water and air in many (though not all) parts of the world. These examples recognise the fundamental importance of multilateralism in bringing member states together to discuss future pathways and implement cross-border legislation. On the other hand, there has been little progress in fighting over-consumption, curbing CO_2 emissions, realising environmental and social justice, creating a circular and distributive economy fuelled by solidarity, and intergenerational and interspecies justice. As a result, humanity is at an existential turning point in terms of addressing the Earth's triple planetary crisis – climate, nature, and pollution (UNEP, 2023). In addition to environmental aspects, this triple crisis recognises the monumental challenges to addressing social and economic dimensions which require deep transformations in how we relate to one another and the environment.

Fifty years on from Stockholm 1972, there have been 28 United Nations Climate Change Conferences (COP), as well as 15 COP biodiversity summits, and three overarching Earth Summits. It is an open question as to the extent to which these international summits have been successful in moving humanity in a direction to which it can eventually live within the limits of what our natural and social environments can sustain.

What has become clear is the importance of education and learning in addressing what we can term a crisis of culture. Rather than creating and supporting cultures that are caring, community, and solidarity-oriented, cultures that enable greed, individualism, exploitation, and commodification are

Stockholm+50 was a major international environmental meeting that took place from 2 to 3 June 2022 in Stockholm, Sweden, in the lead-up to World Environment Day. The theme of the conference was 'Stockholm+50: a healthy planet for the prosperity of all – our responsibility, our opportunity'. Anchored in the Decade of Action, this high-level meeting had the goal of accelerating a transformation that leads to sustainable and green economies, more jobs, and a healthy planet for all – where no one is left behind. An important output of this conference was the People's Environment Narrative (PEN) – a compilation of articles documenting fifty years of efforts and accomplishments in order to safeguard the environment (Strandenaes & Alvarez, 2022). The PEN legacy paper 'Fifty Years of Education and Learning for the Environment and Sustainability', brought the authors of this book together on a collaborative journey which resulted in this text.

Figure 1.2 The Stockholm+50 conference

Image source: Strandenaes and Alvarez (2022)

thriving. In response, there are calls for the types of education and learning that can speak to deeper layers of thinking and question ingrained assumptions about how we live, govern and organise ourselves (Tilbury, 2011). This goes beyond changing individual behaviours and requires us to be socially critical about predicted and preferred futures.

Anchored, as we are, in the UN Decade of Action, a suggested major goal of education and learning is to contribute to accelerating the transformations needed to reach the 17 Sustainable Development Goals (SDGs) of Agenda 2030. This involves, among others, quality education for all, decent work, clean water, health and well-being, zero hunger and no poverty, as well as climate action and strong institutions.

While it is clear that education and learning has a fundamental role to play in establishing a more just relationship with ourselves, other species, and the planet as a whole, 50 years of policy, diplomacy, and governance have not changed much in our schools, universities and non- and informal systems of education and learning (Wals et al., 2022). In this book, we ask the question what needs to be done for education's role to materialise in an effective way. It recognises that our future depends on efforts to transform education and learning so that it can challenge the way we engage with the natural environment and create alternative sustainable futures.

What Difference Has Education Made? Tracking Key Educational Currents and Contributions through the Decades (1970–2020)

The context of the Stockholm 1972 conference provided the initial entry point into international collaboration and agreements in relation to education and engagement in support of a better environment. Over the years, the changing context and shifting narratives on education have shaped our understanding and approaches to learning for the environment. The following themes transverse the development of educational and learning-based responses to environmental and sustainability concerns from the 1970s until the 2020s.

i *The role of education:* In efforts to support education and learning for the environment, a fundamental question is how the role of education has been evolving. Over the decades, we see changes in the underlying educational philosophy and assumptions questioning to what end and how we can best educate for the planet.

ii *Thematic entry points for education:* As social, economic, and ecological contexts evolve and our knowledge expands and deepens, we see shifts in the environmental issues that become the focus of learning and education efforts in this area. This element will track the different thematic entry points for each decade.

iii *Where learning happens:* Learning takes place within the context of relationships among learners, educators, families, communities, and their environments. As we move through the decades, we will highlight these relationships but also how sites of learning for the environment have been extended or evolved over time.

iv *Involvement of stakeholders:* Education involves diverse stakeholders, which have interests in advancing ideas, or investing in education's successful development. Stakeholders affect decision-making, and so we track the shifting influence of diverse stakeholders including international agencies, scientists and environmentalists, psychologists and academics, education experts, NGOs, public and private organisations, community groups, parents, and young people. This book reviews the crucial role they have played in shaping learning for environment and sustainability through the decades.

v *Narratives and paradigms influencing education and environment:* This theme explores how we see the world and how this has evolved over time. We track the assumptions and aspirations underpinning the paradigms and how they shape how we see and engage with education and learning for the environment over the years.

As the chapters move more in-depth to a decade-by-decade description of these educational currents, it is important to remember that elements of all the different trends and narratives described, co-exist today, although some may dominate, while others loom in the shadows or are marginalised. Also, we recognise that there is overlap between the decades, with some important ideas being introduced in one decade, but not having educational implications until later on. Nevertheless, we considered it possible to identify patterns, as well as a number of considerations that transcend the different decades. In Figure 1.3, we present a summary of these trends, as well as the frameworks that have shaped international engagement in education and learning.

This Book

Moving on from this chapter, in which we provided a background and context to the 1972 Stockholm declaration, the following Chapters 2–7 provide a decade-by-decade overview of the last 50 years, with a summary in Chapter 8, exploring the five transversal themes outlined above, as well as international landmark events that have shaped educational responses to environmental challenges.

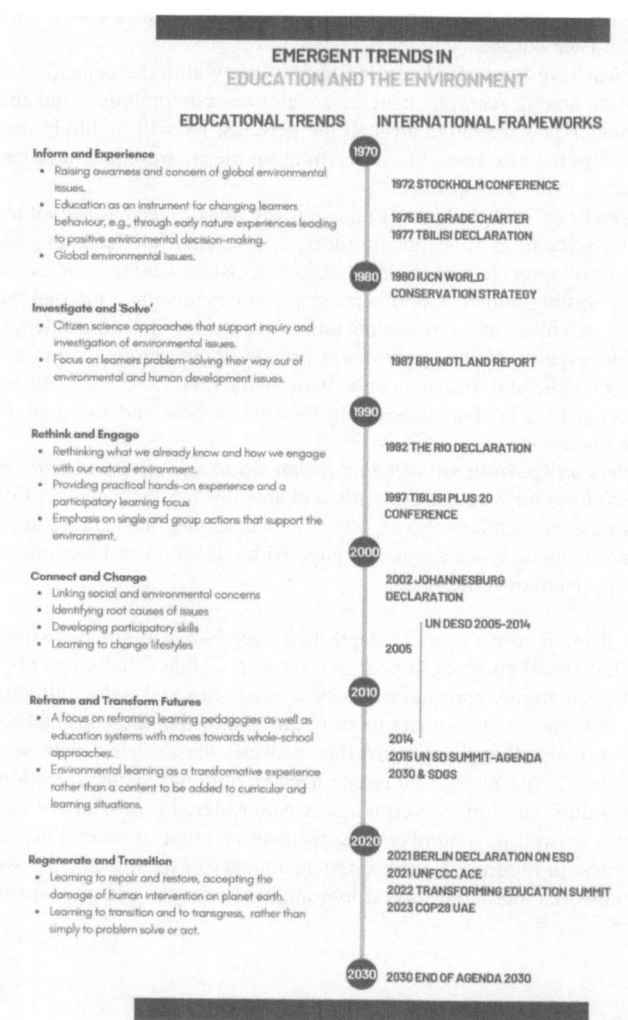

Figure 1.3 Emergent trends in education and the environment (1970–2030)

Chapter 2 explores the decade of the 1970s, whereby education was characterised by the need to *Inform and Experience*, thus increasing awareness in relation to emerging environmental issues. While the 1972 Stockholm declaration cemented the importance of Environmental Education, it was the 1975 Belgrade charter and Tbilisi conference of 1977 that added more substance to its meaning. At this time, education was interpreted by many as a tool for inculcating a love for nature or as an instrument for changing learner behaviour. It was mostly seen as a policy instrument alongside technological development, legislation and financial incentives and used to combat environmental degradation.

Chapter 3 details the 1980s, which saw the educational trend to *Investigate and Solve* through science and technology, but also the need to engage not just experts, but also citizens in solving our way out of problems. At the same time, there was significant questioning from educational circles about the value of an instrumentalist form of learning and the need to understand socio-political root causes of issues. A shift in the educational frames informing environmental learning served to challenge pedagogical processes and extend the interest in Environmental Education across schools, colleges, and universities. Building on the IUCN's world conservation strategy in 1980, this decade saw the publication of 'Our Common Future' in 1987, commonly referred to as the Brundtland report, directing nations towards the goal of sustainable development (WCED, 1987).

Chapter 4 takes a look at the 1990s, which saw a shift towards a *Rethink and Engage* approach to education. An important landmark was the Rio Declaration of 1992, which redefined the issues surrounding Environmental Education identified at Stockholm within the new language of sustainable development. At the same time, the Tbilisi+20 conference in 1997 revealed the divides and tensions resulting from the sustainable development paradigm entering educational approaches. The influence of this discourse was seen in educational circles, where critiques emerged of Environmental Education practices that sought to celebrate and preserve nature. Calls were made to examine, question, and rectify the unequal social relations embedded in contemporary society that lead to the exploitation of the environment. This paradigm shift resulted in curriculum, pedagogical, and thematic changes towards educating for sustainability.

Chapter 5 moves on to the 2000s, which saw a consolidation of the emergent approaches over the last decades towards a *Connect and Change* approach in education. Increasing attention was given to more integrative and emancipatory approaches to education which questioned the root causes of socio-ecological concerns as well as learning to connect peoples' interests and lifestyles with these issues. The Johannesburg declaration in 2002 led to the United Nations Decade of Education (DESD) that took place between 2005 and 2014.

Chapter 6 investigates the 2010s, which saw a convergence of educational streams around the drive to 'Reframe and Transform Futures' towards addressing sustainability concerns. The 17 Sustainable Development Goals (SDGs) of the 2030 Agenda for Sustainable Development officially came into force, following the adoption by world leaders in the 2015 historic UN Summit. At the same time decolonising voices and proposals, often based on alternative development models, offered new ways of teaching and learning, contributing to reframing education towards whole-school approaches and transformative learning experiences.

Chapter 7 finishes off the decades by reviewing the 2020s, as we look towards the goals set by the UN 2030 agenda. Characterised by the educational trend to *Regenerate and Transition*, this chapter notes exciting new strands of education that are (re)emerging in the 2020s. In addition, the limitations to the transmissive, classroom approach to sustainability education have become clear, with increasing moves towards boundary crossing between education in relation to health, climate, citizenship, inclusion, and justice. The Berlin declaration on ESD highlighted the importance of transformative learning for people and the planet, with the United Nations Framework Convention on Climate Change's Action for Climate Empowerment (ACE) strengthening the role of communication and education in the attainment of climate action.

The final chapter, Chapter 8, summarises the key educational themes listed in Figure 1.3 across the five decades. With a focus on policy and governance issues, the chapter frames a set of recommendations to upscale, improve, and connect efforts in education and learning for a better environment. A key message is that Environmental Education, Education for Sustainable Development, learning for sustainability; climate education, global education, or any learning that supports a healthy planet has to have educational value and seek to transform education itself, if we aspire to more sustainable communities.

A final note before we move on: being aware and reflective of past trends in education and the environment can be helpful for a new generation entering this space with a commitment to learning for the planet. These new voices sometimes push for approaches and paradigms that in the past have shown to be ineffective in our quest for a healthier planet. This highlights the importance of familiarising ourselves with the history and lessons learnt regarding the role of education in learning to live within planetary boundaries. The authors recognise the significance of engaging a broader group of stakeholders in the educational process but also of the concern that a lack of context or learning can delay or side-track efforts to realise a better planetary future. It is for this reason that we seek to trigger a dialogue that enhances debates about the *why*, *what*, *where* and *how* of learning for the environment in the quest for more sustainable futures.

References

Botha, L., Griffiths, D., & Prozesky, M. (2021). Epistemological Decolonization through a Relational Knowledge-Making Model. *Africa Today, 67*(4), 51–72.

Botzler, R. G., & Armstrong, S. J. (1993). *Environmental Ethics: Divergence and Convergence*. http://repo.unikadelasalle.ac.id/index.php?p=show_detail&id=1979&keywords=

Carson, R. (1962). *Silent Spring*. Houghton Mifflin.

Clarke, R., & Timberlake, L. (1982). *Stockholm Plus Ten. Promises, Promises? The decade since the 1972 UN Environment Conference*. International Institute for Environment and Development (IIED).

Handl, G. (2012). Declaration of the United Nations Conference on the Human Environment (Stockholm Declaration), 1972 and the Rio Declaration on Environment and Development, 1992. *United Nations Audiovisual Library of International Law, 11*, 6.

Hicks, D. (2012). *Sustainable Schools, Sustainable Futures*. Godalming: WWF. https://www.risingstars-uk.com/media/Rising-Stars/Series%20Images/Voyagers%20Free%20Samples/Sustainable-Schools,-Sustainable-Futures-Contents.pdf

Hoffmann, D. L., Standish, C. D., García-Diez, M., Pettitt, P. B., Milton, J. A., Zilhão, J., Alcolea-González, J. J., Cantalejo-Duarte, P., Collado, H., de Balbín, R., Lorblanchet, M., Ramos-Muñoz, J., Weniger, G.-C., & Pike, A. W. G. (2018). U-Th Dating of Carbonate Crusts Reveals Neandertal Origin of Iberian Cave Art. *Science, 359*(6378), 912–915.

Leinfelder, R. (2013). Assuming Responsibility for the Anthropocene: Challenges and Opportunities in Education. *RCC Perspectives, 3*, 9–28.

Lovejoy, A. (1974). *The Great Chain of Being Cambridge*. Harvard University Press.

Meadows, D. H., Meadows, D. L., Randers, J., & Behrens, W. W. (1972). *The Limits to Growth: A Report for the Club of Rome's Project on the Predicament of Mankind*. Universe Books.

McLeman, R. A., Dupre, J., Berrang Ford, L., Ford, J., Gajewski, K., & Marchildon, G. (2014). What We Learned from the Dust Bowl: Lessons in Science, Policy, and Adaptation. *Population and Environment, 35*(4), 417–440.

Muller, S., Hemming, S., & Rigney, D. (2019). Indigenous Sovereignties: Relational Ontologies and Environmental Management. *Geographical Research, 57*(4), 399–410.

Naess, A. (1990). *Ecology, Community and Lifestyle: Outline of an Ecosophy*. Cambridge University Press.

Najam, A., & Cleveland, C. J. (2005). Energy and Sustainable Development at Global Environmental Summits: An Evolving Agenda. In L. Hens & B. Nath (Eds.), *The World Summit on Sustainable Development: The Johannesburg Conference* (pp. 113–134). Springer Netherlands.

Pepper, D. (1984). *The Roots of Modern Environmentalism*. Routledge.

Schumacher, E. F. (1973). *Small Is Beautiful: Economics as though People Mattered*. Harper & Row.

Selin, H. (2003). *Nature Across Cultures: Views of Nature and the Environment in Non-Western Cultures*. Kluwer Academic Press.

Seyfang, G. (2003). Environmental Mega-Conferences—from Stockholm to Johannesburg and Beyond. *Global Environmental Change: Human and Policy Dimensions, 13*(3), 223–228.

Strandenaes, J. G., & Alvarez, I. (Eds.). (2022). *The People's Environment Narrative (PEN) – 50 Years with UNEP and Civil Society*. UNEP.

Tilbury, D. (2011). *Education for Sustainable Development: An Expert Review of Processes and Learning*. UNESCO.

UNEP (2023). *Keeping the Promise: Annual Report 2023*. UNEP.

Wals, A., Pinar, W., Macintyre, T., Chakraborty, A., Johnson-Mardones, D., Waghid, Y., Tusiime, M., Le Grange, L., LL, Razak, D. A., Accioly, I., Xu, Y., Humphrey, N.,

Iyengar, R., Chaves, M., Herring, E., Vickers, E. A., Santamaria, R. D. P., Korostelina, K. V., & Pherali, T. (2022). Curriculum and Pedagogy in a Changing World. In E. A. Vickers, K. Pugh & L. Gupta (Eds.), *Education and context in Reimagining Education: The International Science and Evidence based Education Assessment [Duraiappah, A.K., Atteveldt]*. UNESCO MGIEP.

WCED (1987). *Our Common Future. World Commission on Environment and Development*. Oxford University Press.

Weston, A. (1999). *An Invitation to Environmental Philosophy*. Oxford University Press.

White, L. (1967). The Historical Roots of Our Ecologic Crisis. *Science, 155*, 1203–1207.

2 Education and the Environment in the 1970s

Inform and Experience

Introduction: The Emergence of Environmental Education

As we start this historical review, it is important to imagine what the world was like leading up to the 1970s and the foundational Stockholm 1972 conference. In 1954, amid the Cold War, the USSR launched a satellite into orbit called the Sputnik, leading to a rush in the Western world to accelerate investment and efforts in science and technology education to compete with the soviets. In parallel, there were rising concerns about rapid population growth, increasing pollution, and the economic growth paradigm that underlined it (Gómez-Baggethun & Naredo, 2015). The struggle between these two conflicting agendas dates back over a century (Pepper, 1984), but it was Rachel Carson's book *Silent Spring*, published in 1962, which raised consciousness of the severity of the environmental impact of human activities (Carson, 1962).

Other influential texts of the time, such as *The Population Bomb*, written by Paul Ehrlich (1968), predicted worldwide famine in the 1970s and 1980s due to overpopulation. The 1972 book, *Limits to Growth*, further warned that the exponential rates of economic and population growth would not be able to be supported by the Earth's resources, and would collapse before the end of this century (Meadows et al., 1972). It was science and technology that took man to space that same year and gave humanity a new vision of the future, but ironically, it was Apollo 17's first colour photo of the Earth from space – called the Blue Marble – that provided a different perspective of the earth, demonstrating our vulnerability and reliance on the natural environment.

During the 1970s, environmentalists, academic writers, and international policy frameworks converged on a key message: quality of life is dependent on the quality of the environment. In term, the quality of the environment is itself dependent on the type and intensity of human activity (Schumacher, 1973). Major documents at the time restated the nature of this interdependence and sought to develop a social consciousness underpinned by our treatment of the

DOI: 10.4324/9781003467007-2

planet. However, rather than reframing mindsets and deeply questioning social priorities or daily life choices and their impact on natural systems, efforts remained focused on developing positive relationships with nature, with the outdoors, natural science, and wilderness education becoming popular frames for education (Hungerford, 2009).

Critiques of mainstream economics of the 1970s began to emerge with the work of E.F. Schumacher, *Small Is Beautiful: A Study of Economics as If People Mattered* (1973). Schumacher argued that the modern economy is unsustainable and made a case for education as the greatest resource, philosophically highlighting through the term 'small is beautiful' the type and scale of changes necessary to address environmental challenges. David Pepper, author of the book *The Roots of Modern Environmentalism (1984)*, highlighted the importance of Schumacher's work in demonstrating how our value systems influence what we consider the roles of work and education to be, a debate which continues to this day. During this time, Environmental Education began to slowly emerge as a concept, with William Stapp publishing *The Concept of Environmental Education* in the first issue of *The Journal of Environmental Education* (1969). This laid the academic platform for environmental learning and engaging interest in this area for many years to come.

Environmental Education Emerges as an International Policy Commitment

The 1970s also saw increasing interest by Western organisations such as the United Nations in the field of learning. The Swedish delegation of the United Nations led the drive to acknowledge that environmental issues were affecting all peoples, regardless of race, socio-economic standing, and both developed and developing countries, leading to the 1972 Stockholm conference. This context led to an international policy commitment to education at the international level, first within the IUCN and shortly thereafter in the UN. The concept of Environmental Education was first formalised in policy and governance circles by the International Union for the Conservation of Nature and Natural Resources (IUCN), in 1970 at a meeting in Nevada, USA. At that meeting, Environmental Education was defined as:

> A process of recognising values and classifying concepts in order to develop skills and attitudes necessary to understand and appreciate the inter-relatedness among man, his culture and his biophysical surroundings. Environmental Education also entails practice in decision-making and self-formulating of a code of behaviour about issues concerning environmental quality.

(IUCN, 1970, p. 11)

Principle 19 of the 1972 Stockholm declaration cemented this through demonstrating the important role of education in addressing environmental challenges, as well as recognising the scale of the response needed, from the local to the global (see Figure 1.1 in Chapter 1). While the 1972 declaration was instrumental in establishing a status for the new area of learning called Environmental Education, and providing broad policy goals and objectives, it did not provide detailed normative positions (Handl, 2012). What followed, in 1975, was a global framework for education in the form of the Belgrade Charter on Environmental Education (UNESCO, 1975), which gave form to what was outlined in Principle 19. This declaration stated that Environmental Education constitutes a comprehensive lifelong education responsive to changes in a rapidly changing world, with the goal of developing a world population that is aware of, and concerned about, the environment and its associated problems. The UNESCO-UNEP Tbilisi conference on Environmental Education of 1977 updated, clarified, and expanded the Stockholm declaration guided by the Belgrade Charter (see Figure 2.1). The Tbilisi Declaration was the first major international policy document pointing to the importance of changing prevailing growth and expansion-centred economic logic through education (UNESCO-UNEP, 1977).

As noted by Le Grange and Reddy (2007), the Tbilisi conference was important for stating that Environmental Education should consider the environment in its totality, including the interactions between social and ecological dimensions. This required an interdisciplinary approach to learning, where learners should be active participants in planning their own learning experiences.

The Tbilisi conference proposed new goals, objectives, characteristics, and guiding principles of education concerned with the environment; restating the importance of promoting awareness, knowledge, attitudes, skills, and participation, as well as positioning education as a continuous life-long learning process. The following are the goals of Environmental Education, as stated in the Tbilisi conference, highlighting the focus on awareness raising and environmental behaviour (UNESCO, 1978, pp. 26–27):

a   to foster clear awareness of, and concern about, economic, social, political, and ecological interdependence in urban and rural areas.

b   to provide every person with opportunities to acquire the knowledge, values, attitudes, commitment, and skills needed to protect and improve the environment.

c   to create new patterns of behaviour of individuals, groups, and society as a whole towards the environment.

Figure 2.1 The goals of Environmental Education at the Tbilisi conference

From Awareness Raising to Behaviour Change

Earth Day, first celebrated on April 22, 1970, has played a pivotal role in catalysing Environmental Education. Millions have participated in events, rallies, and educational programs, emphasising the public's appetite for environmental knowledge (O'Riordan et al., 1995). Initially, most countries responded with a series of resources and learning objectives that sought to raise awareness of the global issues that challenged planetary health. This approach, often referred to as education *about* the environment, in its simplest form, sought to heighten awareness of environmental concerns (Jickling & Spork, 1998). This was based on the now questionable assumption that the fear arising from environmental threats and the knowledge of how ecosystems worked would trigger an action response from the learner. Over the years, it has been recognised that education *about* the environment can improve environmental literacy, but will have a limited impact by itself in addressing the environmental situation (Crompton, 2008). Instead, this approach should be taught alongside other learning activities that clarify values and develop systemic thinking and action competence of the learner (Bianchi et al., 2022; Tilbury & Wortman, 2004).

The 1970s also saw the US dominating Environmental Education discourse with its focus on education as an instrument to change environmental behaviour (see Trent, 1983). This approach sees the need to 'correct' people's actions in order to limit environmental damage caused by humans (see Figure 2.2). Strategies are used to modify learners' behaviours, with children and adults often being subjected to pre- and post-instruction tests. The educator predetermines what is considered a 'good' or 'bad' behaviour and seeks to directly influence how learners respond. These strategies were 'infused' into teacher education programmes also with environmental content and knowledge added into existing or planned courses for what was then termed as pre-service teachers (Hungerford et al., 1988). Many educationists found themselves reluctant to engage with Environmental Education as a tool to change behaviour as they considered the task of education more in the realm of personal development, capacity-building and unfolding human potential and not one of fostering certain behaviours. Research has since provided evidence of a complex relationship between knowledge and behaviour and the value offering choice and possibilities for the learner (Scoullos & Malotidi, 2004). Later, Jensen and Schnack (1997) explained how rather than modifying behaviour, the challenge for guiding people is that of helping them discover for themselves the changes which are most meaningful to them and helping them develop the competence to create change.

The Nature Narrative and Outdoor Education

Alongside the pro-environmental behaviour movements, there was also increasing support for the *nature narrative*, with an assumed correlation between developing a connection with nature at an early age in life, and making positive

This decade saw environmental campaigns in schools using slogans like *'Give a hoot! Don't pollute'* and *'If you don't recycle, you're throwing it all away.'* The assumption was that environmental concerns such as littering should be addressed through developing precise behavioural practices for motivating groups of individuals towards pro-environmental behaviour. Research at the time shows how anti-litter programs at schools, based on feedback to schools on cleanliness of school-yards, and activities such as school movies contingent on clean yards, were shown to be effective in reducing litter at schools (Gendrich et al., 1982). However, there was little evidence that the desired environmental behaviour lasted or transferred to positive environmental attitudes or actions as was anticipated at the time (Clayton, 2012).

Figure 2.2 Education as an instrument to promote pro-environmental attitudes and behaviours

Image source: https://www.si.edu/object/poster-give-hoot-dont-pollute%3Anmah_529340

This content is made available under the Creative Commons Zero (CC0) license

environmental decision-making later in life (Tanner, 1980). This view was of great interest to those who were dedicated to protecting natural habitats and working in national parks or field centres. The voices of this group had gained in strength thanks to recently established professional associations and membership bodies. For example, the National Wildlife Federation (NWF) in the US, distributed more than 40,000 Wildlife Week education kits free to schools around the country, and published a series of booklets about the environment for children seeking to engender a sense of wonder in young minds. This perspective assumed a linear relationship between knowledge, awareness, and behaviour and greatly influenced how environmental learning and education activities were assessed or evaluated (Tilbury, 1993). The strength of this lobby was such that it was not until the late 1990s that this educational model was properly questioned, with research demonstrating the problematic nature of this relationship, as there is not often a direct correlation between the variables (Hart & Nolan, 1999; Kollmuss & Agyeman, 2002; Scoullos & Malotidi, 2004). However, the nature narrative is so deeply rooted that decades later the relationship between knowledge, awareness, and behaviour is still a hotly debated topic.

In the 1970s, various non-Western cultures and regions were also engaged in alternative approaches to Environmental Education, drawing on their unique perspectives and traditional knowledge. While specific texts may not be as widely recognised, several movements and initiatives reflected non-Western philosophies on environmental stewardship and education during this period. Examples were emerging areas such as Chinese Environmental Philosophy, Buddhist Environmental Ethics, and Gandhian environmentalism. A representative example comes from Maori Environmental knowledge in New Zealand, where the term *tangata whenua* – people of the land – represents a woven tapestry of narratives on the environment which accumulated knowledge and deepened understanding of people and land. These worldviews often embraced ecological messages and environmental ethics which developed a close affinity with their environments. It supported a worldview underpinned by a strong sense of *Kaitiakitanga* – custodianship – a belief that the environment should be maintained in a fit state for future generations (PCE, 2004). We can see this sentiment emerge in later decades through the Brundtland report's future-oriented definition of sustainable development.

The Importance of Socio-Economic Issues

Towards the end of the decade, there was growing international grassroots concern over the ability of natural scientists to address environmental concerns (Tilbury, 1993). It saw a paradigm shift away from a predominantly natural science framework to one that began to include human and social science interpretations of environmental issues (Robottom, 1987; Williams, 1985). This shift is visible in the 1977 Tbilisi declaration, which highlighted the need to understand the complex relations between socioeconomic development and the improvement of the environment (Fensham, 1978; Hungerford, 2009). Disinger (1986) emphasises that this represented a natural evolution in environmental learning and education in terms of interactions of science and technology with society, and the 'environment of concerns', including the human environment. The Tbilisi declaration and Belgrade charter also begin to question mainstream economics: 'Policies aimed at maximising economic output without regard to its consequences on society and on the resources available for improving the quality of life must be questioned' (UNESCO, 1975, p. 2). While the importance of socio-economic issues emerged in the 1970s, it did not gain significant traction until the 1980s (see Chapter 3).

At the same time, critical approaches to education were beginning to emerge, such as the critical pedagogy of Brazilian educator and philosopher Paulo Freire, who brought a social liberating approach that would lead to a conscientisation, and an awareness of communities living in poverty, who are illiterate and lack the power to improve their livelihoods (Freire, 1970). These power discourses existed in pockets and had little influence in the practice of environmental learning and education in these early years. However, they gained in importance over later decades, becoming a critical component of learning *for* the environment in eco-pedagogical approaches that place a strong emphasis on social and environmental justice (Kahn, 2008, 2010).

Research Journals and Their Influence on Policy and Practice

Since the start of the first journal in the field, the *Journal of Environmental Education* (JEE) in 1969, the academic field of Environmental Education has been characterised by conceptual papers about the meaning of Environmental Education and empirical analytical studies of its impact on learners in terms of their knowledge, understanding, attitudes, and behaviour. The journal provided a podium primarily for researchers in North America and was dominated by universities that were leading in Environmental Education, like the University of Michigan (Stapp), Ohio State (Disinger), and Southern Illinois University (Hungerford and Volk).

Over the years, the number of journals has expanded with the *International Journal of Environmental Education and Information*, and *International Research in Geographical and Environmental Education* first, as well as regionally, such as *The Australian Journal of EE*, *The Southern African Journal of EE* and the *Canadian Journal of EE*. In his reflection on the first 25 years of the journal *Environmental Education Research* (EER), Bill Scott who together with Chris Oulton was the founding editor of EER observes about the early research journals such as JEE: 'because of its aims, scope, and editorial board membership, [the journal] had such a tightly-framed view of what counted as knowledge and what merited publication as good research, that a significant amount of the field's output was prevented from being published there' (2020, p. 1681).

Moving from the 1970s into the 1980s

The 1970s saw an intergovernmental policy commitment to Environmental Education which laid the groundwork for a global shift towards integrating it into formal and informal learning settings. Following the influence of the 1972 Stockholm meeting, this decade saw environmental learning defined by a nature-based narrative that would aspire to rebuild our relationship with the natural environment. At the same time, the decade was shaped by a dominant and perhaps contradictory view, that science and technology would solve our environmental plight. While the decade's key contribution was arguably its ability to instil the notion that quality of life is dependent on the quality of the environment, a simmering tension arising in the late 1970s would underlie the following decades as the role of education in addressing structures and practices that exploit the planet came into question.

Specifically, educational policy and practice related to environmental learning were primarily focused on raising awareness about environmental issues, establishing connections with the natural environment, and developing technical and scientific responses to global environmental challenges (Gough, 2006; Hungerford, 2009). By the end of the 1970s, education belonged to a whole basket of instruments and tools for changing environmental behaviour which included technological development, legislation, and financial incentives, whereby education was seen as one instrument that could be used to combat environmental degradation. It is important to note that some acute observers such as Robert Stevenson have pointed to a 'pronounced discrepancy' (2007, p. 139) that existed between the intentions to engage learners with the environmental crisis and an emphasis on the acquisition of environmental knowledge and awareness in school programs. Fensham (1978), in a similar vein, argued that environmental learning and education was frequently misunderstood or misinterpreted, and that its core intention was to not raise awareness but to directly address the mindset shifts, economic models, and social engagement levels required to get to the root causes of environmental degradation and shift us towards more environmentally sustainable models of development. However, pedagogically speaking, learning opportunities remained expert driven, teacher-centred, and involved exploratory elements restricted to natural environments. This meant that formal environmental learning took place at visitor centres, in school gardens and playgrounds, in field centres, wilderness areas, and sometimes in Environmental Education centres that were being formed at that time. In higher education, new course offerings on environmental studies and environmental science began to appear See (Figure 2.3) for a summary of education and the environment in the 1970s.

The following chapter details the decade of the 1980s, represented with a focus on science and technology, targeting individual values and behaviours but also saw some significant questioning, from educational circles, about the

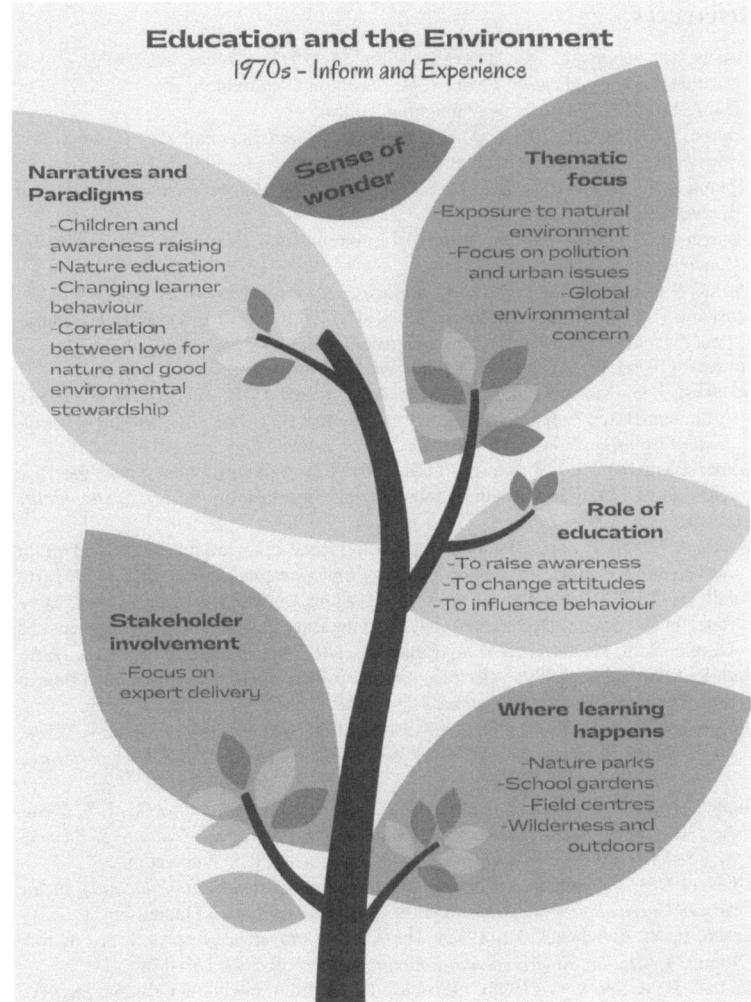

Education and the Environment
1970s - Inform and Experience

Narratives and Paradigms
-Children and awareness raising
-Nature education
-Changing learner behaviour
-Correlation between love for nature and good environmental stewardship

Sense of wonder

Thematic focus
-Exposure to natural environment
-Focus on pollution and urban issues
-Global environmental concern

Role of education
-To raise awareness
-To change attitudes
-To influence behaviour

Stakeholder involvement
-Focus on expert delivery

Where learning happens
-Nature parks
-School gardens
-Field centres
-Wilderness and outdoors

Figure 2.3 Summary of education and the environment in the 1970s

value of this learning. As the decade progressed, a shift was witnessed in the narratives with the realisation that environmental problems were no longer a 'clean up' problem, solved by increasing awareness to the issues and technological and scientific solutions. Instead, an understanding of socio-political and economic contexts was seen as vital to get to the root causes.

References

Bianchi, G., Pisiotis, U., & Cabrera Giraldez, M. (2022). *GreenComp The European Sustainability Competence Framework*. European Commission.

Carson, R. (1962). *Silent Spring*. Houghton Mifflin.

Clayton, S. D. (2012). *The Oxford Handbook of Environmental and Conservation Psychology*. Oxford University Press.

Crompton, T. (2008). *Weathercocks and Signposts: The Environment Movement at a Crossroads*. WWF.

Disinger, J. F. (1986). Current Trends in Environmental Education. *The Journal of Environmental Education, 17*(2), 1–3.

Ehrlich, P. R. (1968). *The Population Bomb*. Yale University Press.

Fensham, P. J. (1978). Stockholm to Tbilisi–The Evolution of Environmental Education. *Prospects: Quarterly Review of Education, 8*, 446–455.

Freire, P. (1970). *Pedagogy of the Oppressed*. Continuum.

Gendrich, J. G., McNees, M. P., Schnelle, J. F., Beegle, G. P., & Clark, H. B. (1982). A Student-Based Anti-Litter Program for Elementary Schools. *Education & Treatment of Children, 5*(4), 321–335.

Gómez-Baggethun, E., & Naredo, J. M. (2015). In Search of Lost Time: the Rise and Fall of Limits to Growth in International Sustainability Policy. *Sustainability Science, 10*(3), 385–395.

Gough, A. (2006). A Long, Winding (and Rocky) Road to Environmental Education for Sustainability in 2006. *Australian Journal of Environmental Education, 22*(1), 71–76.

Handl, G. (2012). Declaration of the United Nations Conference on the Human Environment (Stockholm Declaration), 1972 and the Rio Declaration on Environment and Development, 1992. *United Nations Audiovisual Library of International Law, 11*, 6.

Hart, P., & Nolan, K. (1999). A Critical Analysis of Research in Environmental Education. *Studies in Science Education, 34*(1), 1–69.

Hungerford, H. R. (2009). Environmental Education (EE) for the 21st Century: Where Have We Been? Where Are We Now? Where Are We Headed? *The Journal of Environmental Education, 41*(1), 1–6.

Hungerford, H. R., Volk, T. L., Bluhm, W. J., Dixon, B. G., Marcinkowski, T. J., & Sia, A. P. C. (1988). *An Environmental Education Approach to the Training of Elementary Teachers: A Teacher Education Programme*. UNESCO Publications.

IUCN (1970). *International Working Meeting on Environmental Education in the School Curriculum*. Foresta Institute for Ocean and Mountain Studies.

Jensen, B. B., & Schnack, K. (1997). The Action Competence Approach in Environmental Education. *Environmental Education Research, 3*(2), 163–178.

Jickling, B., & Spork, H. (1998). Education for the Environment: a Critique. *Environmental Education Research, 4*(3), 309–327.

Kahn, R. (2008). From Education for Sustainable Development to Ecopedagogy: Sustaining Capitalism or Sustaining Life? *Green Theory & Praxis The Journal of Ecopedagogy, 4*(1), 1–14.

Kahn, R. (2010). *Critical Pedagogy, Ecoliteracy, & Planetary Crisis: The Ecopedagogy Movement*. Peter Lang.

Kollmuss, A., & Agyeman, J. (2002). Mind the Gap: Why Do People Act Environmentally and What Are the Barriers to Pro-Environmental Behavior? *Environmental Education Research, 8*(3), 239–260.

Meadcws, D., Meadows, D., Randers, J., & Behrens, W. (1972). *The Limits to Growth: A Report to the Club of Rome (1972)*. Earth Island Ltd.

O'Riordan, T., Clark, W. C., Kates, R. W., & McGowan, A. (1995). The Legacy of Earth Day: Reflections at a Turning Point. *Environment: Science and Policy for Sustainable Development, 37*(3), 7–37.

PCE (2004). *See Change: Learning and Education for Sustainability*. Parlimentary Commissioner for the Environment.

Pepper, D. (1984). *The Roots of Modern Environmentalism*. Routledge.

Le Grange, & Reddy (2007). Think Piece. Learning of Environment (s) and Environment (s) of Learning. *Southern African Journal of Environmental Education*. https://www.ajol.info/index.php/sajee/article/view/122741

Robottom, I. (1987). Contestation and Consensus in Environmental Education. *Curriculum Perspectives, 7*(1), 23–27.

Schumacher, E. F. (1973). *Small Is Beautiful: Economics as though People Mattered*. Harper & Row.

Scott, W. (2020). 25 Years on: Looking Back at Environmental Education Research. *Environmental Education Research, 26*(12), 1681–1689.

Scoulles, & Malotidi. (2004). *Handbook on methods used in Environmental Education and Education for Sustainable Development*. Mio-ECSDE.

Stapp, W. B. (1969). The Concept of Environmental Education. *The Journal of Environmental Education, 1*(3), 31–36.

Stevenson, R. B. (2007). Schooling and Environmental Education: Contradictions in Purpose and Practice. *Environmental Education Research, 13*(2), 139–153.

Tanner, T. (1980). Significant Life Experiences: A New Research Area in Environmental Education. *The Journal of Environmental Education, 11*(4), 20–24.

Tilbury, D. (1993). Environmental Education: Developing a Model for Initial Teacher Education *(Doctoral dissertation)*. University of Cambridge.

Tilbury, D., & Wortman, D. (2004). *Engaging People in Sustainability*. IUCN.

Trent, J. H. (1983). Environmental Education in Our Schools During the 1970s. *The Journal of Environmental Education, 15*(1), 11–15.

UNESCO (1975). *Belgrade Charter on Environmental Education*. UNESCO.

UNESCO (1978). *Intergovernmental Conference on Environmental Education: Tbilisi Declaration*. UNESCO.

UNESCO-UNEP (1977). The Tbilisi Declaration. *Intergovernmental Conference on Environmental Education*, 14–26. UNESCO.

Williams (1985). *Environmental Education and Teacher Education Project 1984-1987*. World Wildlife Fund-Unpublished.

3 Education and the Environment in the 1980s

Investigate and Solve

Introduction: Scientific Problem-Solving and the Rise of the NGOs

The 1980s brought new and more engaging visions of environmental learning, with the key Stockholm messages taking root in the education world. However, the decade also drew out social and educational tensions, many of which were rooted in the 1972 Stockholm Declaration. The 1980s placed importance on science education, which had become the dominant pathway for young people to learn about the environment in school (Lucas, 1980), overtaking the more outdoor learning and nature approaches seen in the 1970s. In the earlier years of the decade, environmental issues were framed as a scientific problem to be solved and not just as a science to be understood. This influence saw a more linear approach to learning *about* the environment, how environmental problems were identified, and how individuals and small groups were expected to solve the big problems the world was facing. The approach was still predominantly cognitive, as factual knowledge was considered necessary for more appropriate environmental decision-making (Sauvé, 2005). However, there was an extended responsibility to citizens with the science remit no longer limited to qualified professionals (see Figure 3.1).

Responding to ecosystem decline and social stresses, international NGOs claimed their place in the social dialogues of the 1980s (Caldwell, 1988). This provided a critical counterweight to dominant trends in the global political economy, at all levels, from the local to the global (Finger & Princen, 2013). For example, NGOs have been widely credited with performing an instrumental role in pushing for the 1987 Montreal Protocol on Substances that deplete the Ozone Layer, owing largely due to UNEP's policy to involve non-state actors, whereby NGOs participated directly in the preparatory and actual negotiations (Finger & Princen, 2013). In countries such as Brazil, environmental NGOs played an important role in the process of expanding non-formal environmental learning, catalysing governmental initiatives, and providing support to private organisations working on environmental learning initiatives (Tristão & Tristão, 2016).

DOI: 10.4324/9781003467007-3

'Riverwatch' and 'Adopt-Stream' in North America and, in Europe, the 'Blue Flag program' and Ireland-based 'Coastwatch' are examples of citizenship science programmes designed to monitor and improve the health of the coastline. It involves volunteers from all walks of life checking their chosen 500m stretch of coast (survey unit) once around low tide, and jotting observations down on the survey questionnaire while on the shore. This citizen science work is often augmented with water tests. Data is then collected and pooled to provide a snapshot of the environmental state of the coastline areas surveyed at that time. Such programs demonstrate the increasing engagement of citizens in scientific fields in the 1980s, areas normally reserved for experts.

Figure 3.1 Citizenship, engagement, and science

Image source: The Valley Reporter (2022)

This decade also saw the first document that formally addressed the issue of teacher professional development on matters of the environment (Wilke et al., 1987). Rather than problematise and identify tensions between local–global, human–environment and teacher–student driven approaches, this document mapped out specific content as an instructional manual. This framework approach was to be challenged very shortly afterwards in the early 1990s.

At the higher education level, there continued to be only specialist offerings in environmental education focused on the training of specialists in environmental engineering; environmental sciences and in the mid-1980s, in Russia, for example, 'Environmental Protection and Rational Use of Natural Resources' was introduced at technical universities developing graduates with relevant expertise in this area (Shutaleva et al., 2020).

Beyond Science

Over the 1980s, NGOs increasingly influenced educational responses to the environment as they pushed the notion that science was not enough,

arguing that human activity, which was causing environmental degradation, needed to be addressed through a people's perspective. They also
questioned the increasing influence of environmental experts on what was
perceived as a social problem. This emerging trend can be illustrated in a
series of teacher's activity guidebooks developed by WWF-UK, entitled
the Global Environmental Education Programme. The *What We Consume*
module, written by John Huckle (1988), positioned environmental learning within the economic, political, social, and philosophical structures that
direct human activity and influenced social values (Martin, 1985). Greig
et al.'s (1987) *Earthrights: Education as If the Planet Really Mattered,* was
another seminal document of the 1980s that captured the interest of educators who had previously failed to connect with the purely techno-scientific
thrust of the environmental agenda (Tilbury, 1993). The Earthrights document also supported boundary crossing and made the case for aligning environmental concerns within the broader umbrella of adjectival education
such as futures education, global education, health education, citizenship
education, and multicultural education. This framing gained significant
traction in the following decade when sustainable development became a
key discourse.

Robottom (1987) brought together the work of key Australian influencers,
who called for broadening the scope of environmental issues taught in schools
and for exploring issues through socially -critical frames. The group advocated for a shift away from traditional approaches to learning *in* and *about*
the environment, which were knowledge and science centred, to a focus on
questioning the root causes and socio-cultural frames that led to the exploitation of the environment (Tilbury et al., 2005). Grounding this work was
the Geography Teachers Association of Victoria, which published New Wave
Geography, a textbook for teachers that presented environmental education
as a social concern. It encouraged learners to campaign for a future where
people took personal responsibility for the environment but also called for
governments to protect biodiversity and address land degradation problems
(Geography Teachers Association of Victoria, 1988).

During the late 80s, parallel and in connection to this broadening of the
scope, scholars started to call for greater diversity in Environmental Education research. Up until that point, most research in the field was of an empirical analytical 'positivist' nature, ignoring interpretative 'hermeneutical'
and socially critical strands of research (Mrazek, 1993). Alongside this call
to refresh how we perceived environmental concerns, the role of education in
addressing them and the way research on Environmental Education was con-
ceived of, there were attempts to align and connect Environmental Education
with social justice and human rights perspectives. The latter brought a focus
on the political and socio-cultural threads that helped get to the bottom of why
it would prove difficult to address the root causes of environmental issues
through science or behaviour change models.

Research Influences on Environmental and Sustainability Education

Research has played a key role in informing and advancing policy and practice in Environmental Education from the early 1980s. The Carbondale Group, from the University of Southern Illinois, was particularly influential in determining the course of environmental education research in its initial years with their work informing UNESCO-UNEP's International Environmental Education Programme (IEEP). These documents were seminal in that they were the first to consolidate research thinking around the field; they were also available in other languages including Spanish, French setting the agenda for research priorities and practice around the globe (Robottom & Hart, 1993). The training of these researchers tinged their view of what counted as valid research and pushed for quantitative inquiry explaining the dominance of experimental and correlational research in the early years. Iozzi (1981) reported that 90–92 per cent of the research undertaken in Environmental Education was primarily quantitative and sought to legitimate scientific knowledge. Research approaches and tools such as psychometrics, behaviour analysis, natural resource management, human ecology, and statistics figured significantly with Environmental Education essentially becoming a branch of science education (Tilbury & Walford, 1996). A decade later, Gough (1993) and Robottom and Hart (1993) pointed to how these entrenched views and promotion of behaviourist, positivist, instrumentalist, and deterministic views of education were critical of the researchers imposing their social values on the research and in ways that disempowered the teachers and educators. Tilbury and Walford (1996) saw these as a flawed form of inquiry contradicting the interdisciplinary and socially critical perspectives embedded in learning processes for the environment and damaging the progress of Environmental Education in practice.

Questioning the Educational Benefits

A further response was a kickback from education specialists concerned with the lack of educational frames in education and learning. A dominant concern of the 1980s was the emphasis on solving environmental problems 'for the good of the planet' rather than for the benefit of learners. Many educationalists in schools, colleges, and higher education were suspicious of the lack of clear educational frames or outcomes. A few experts labelled Environmental Education as instrumentalist in nature and its activities as indoctrination (Hart & Stevenson, 2019). Such experts rejected a values-inculcation approach as well as the implantation of knowledge which lacked any form of interrogation or co-creation process. This can also be seen in the attempts to export a form of Environmental Education to non-western countries, whereby the assumption of the universality of such education was not aligned with the socio-political context of schooling (Vulliamy, 1987). Rather than knowledge-based studies, where learning the skills needed to move on to the next academic year were taught, the environmental experiences of children in developing countries could instead be used for issue-based studies with the assumption that children would be better equipped to make decisions and participate in community action on issues relevant to their lives (Knamiller, 1983). Citing the worldviews of Aboriginal Australians, based on a sense of kinship with the Earth, Gough notes the overall instrumentalist approach to western Environmental Education, and the limitations the term Environmental Education has on representing different relationships to place and nature (Gough, 1990).

Fritjof Capra's *The Turning Point: Science, Society, and the Rising Culture* (1983) provided an alternative pathway seeking to break silos and parallel streams by advocating a more organic and systems approach to understanding reality. In educational terms, this meant moving away from a reductionist and fragmented worldview, where learners are passive vessels, to where teachers act to facilitate the development of students, and where not just the natural environment is important, but also the social and cultural connections. The work of Bill Stapp and his students at the University of Michigan, working with inner-city Detroit Middle Schools engaging youth in their local environment and working on self-identified issues of concern, illustrates this new trend towards student-centred learning as well as boundary crossing between the social and the environmental (Wals et al., 1990). Such trends promoted cross-curriculum and more participatory forms of learning, which explore the linkages between society and environment, global and local issues as well as politics and power from an intercultural perspective. These innovative practices took time to take root in practice, but their influence can still be seen in many schools' national curricula decades later.

Sustainable Development

The later part of the 1980s saw other shifts in the way environmental issues were framed. The 'Our Common Future' report, also referred to as the Brundtland report, directed nations towards the goal of sustainable development, highlighting the moral issue of how today's actions affect future generations (WCED, 1987). It presented environmental problems as not just ecological in nature, but also with social, cultural, and economic dimensions, bringing into focus the now ubiquitous concept of Sustainable Development that had been introduced a few years earlier in the IUCN's World Conservation Strategy (IUCN, 1982). Compared to environmental reports of the 1970s, such as the 1972 Stockholm report, in which the role of economic growth was seen as a growing concern in terms of ecological decline, the Brundtland report instead presented growth as the solution to social and environmental problems (Gómez-Baggethun & Naredo, 2015).

The Brundtland report was thus seen as reformist, rather than transformative of current social or economic systems. The report anthropocentrically treated the natural environment as part of policy options, with the need for technological and economic tools, and advocated a shift in individual and industry behaviour towards a more sustainable road of economic development (Fien & Tilbury, 2002). This would generate increasing tensions and extend debates in educational and environmental circles for decades to come. See Figure 3.2 for a summary of education and the environment in the 1980s.

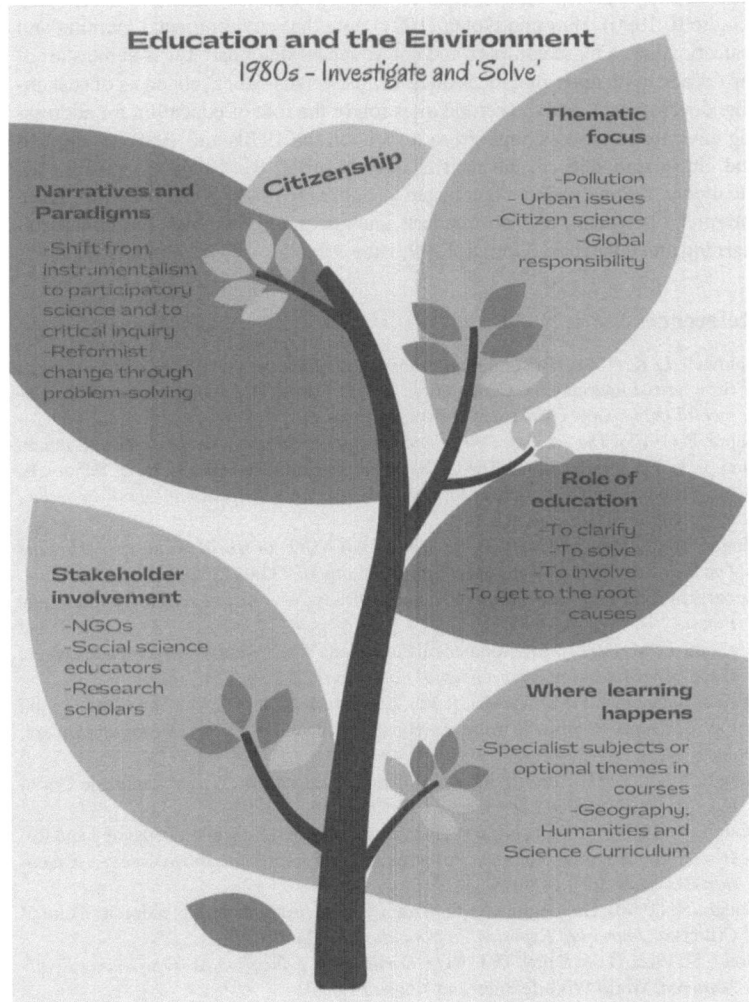

Figure 3.2 Summary of education and the environment in the 1980s

Moving from the 1980s into the 1990s

It took 10 years for the key Stockholm messages to take root in the education world. However, although environmental learning of the preceding decades led to a more environmentally aware population at the end of the 1980s, it was argued that people still lacked the necessary understanding of the roots of sustainability problems and specifically what actions they could or should take (Gigliotti, 1990). Hart and Nolan (1999) note that environmental learning and education was a more complex and controversial field than it was at the start of the decade, with diverging ideas on reformist versus radical concepts of sustainable development, and the central question of the role of education for addressing environmental concerns. From a focus in the 1970s and 1980s on applied and citizenship science, the next chapter covering the 1990s will detail how the theme 'rethink and engage' began to dominate practice and called for more interpretive, critical, and postmodern lines of inquiry through environmental learning and education (Gough, 1999; Hart & Nolan, 1999; Palmer, 2002).

References

Caldwell, L. K. (1988). Beyond Environmental Diplomacy: The Changing Institutional Structure of International Cooperation. In J. E. Carroll (Ed.), *International Environmental Diplomacy*. Cambridge University Press.

Capra, F. (1983). *The Turning Point: Science, Society, and the Rising Culture*. Bantam.

Fien, J., & Tilbury, D. (2002). The Global Challenge of Sustainability. In D. Tilbury, R. Stevenson, J. Fien & D. Schreuder (Eds.), *Education and Sustainability: Responding to the Global Challenge*. IUCN.

Finger, M., & Princen, T. (2013). *Environmental NGOs in World Politics: Linking the Local and the Global*. Routledge. https://doi.org/10.4324/9780203429037

Geography Teachers Association of Victoria (1988). *New Wave Geography*. Jacaranda Press.

Gigliotti, L. M. (1990). Environmental Education: What Went Wrong? What Can Be Done? *The Journal of Environmental Education, 22*(1), 9–12.

Gómez-Baggethun, E., & Naredo, J. M. (2015). In Search of Lost Time: the Rise and Fall of Limits to Growth in International Sustainability Policy. *Sustainability Science, 10*(3), 385–395.

Gough, A. G. (1993). Globalizing Environmental Education: What's Language Got to Do with it? *Journal of Experiential Education, 16*(3), 32–39.

Gough, A. (1999). Recognising Women in Environmental Education Pedagogy and Research: Toward an Ecofeminist Poststructuralist Perspective. *Environmental Education Research, 5*(2), 143–161.

Gough, N. (1990). Healing the Earth Within Us: Environmental Education as Cultural Criticism. *Journal of Experiential Education, 13*(3), 12–17.

Greig, S., Pike, G., & Selby, D. (1987). *Earthrights: Education as If the Planet Really Mattered*. World Wildlife Fund and Kogan Page.

Hart, P., & Nolan, K. (1999). A Critical Analysis of Research in Environmental Education. *Studies in Science Education, 34*(1), 1–69.

Hart, P., & Stevenson, B. (2019). Preamble [*Journal of Environmental Education*, 50, 4–6]. *The Journal of Environmental Education, 50*, 239–240.

Huckle (1988). *What We Consume: The Teachers' Handbook*. World Wide Fund for Nature (UK) and Richmond.

Iozzi, L. A. (1981). *Research in Environmental Education 1971–1980*. ERIC.

Knamiller, G. (1983). Environmental Education for Relevance in Developing Countries. *The Environmentalist, 3*(3), 173–179.

Lucas, A. M. (1980). Science and Environmental Education: Pious Hopes, Self Praise and Disciplinary Chauvinism. *Studies in Science Education, 7*(1), 1–26.

Martin, P. (1985). The WWF-UK's Education Response to the World Conservation Strategy. *Review of Environmental Education Developments, 13*(2), 11–14.

Mrazek, R. (Ed.) (1993). *Alternative Paradigms in Environmental Education Research*. North American Association for Environmental Education.

Palmer, J. (2002). *Environmental Education in the 21st Century: Theory, Practice, Progress and Promise*. Routledge.

Robottom, I. (Ed.). (1987). *Environmental Education: Practice and Possibility*. Deakin University Press.

Robottom, I. M., & Hart, E. P. (1993). *Research in Environmental Education: Engaging the Debate*. Deakin University.

Sauvé, L. (2005). Currents in Environmental Education: Mapping a Complex and Evolving Pedagogical Field. *Canadian Journal of Environmental Education, 10*(1), 11–37.

Shutaleva, A., Nikonova, Z., Savchenko, I., & Martyushev, N. (2020). Environmental Education for Sustainable Development in Russia. *Sustainability: Science Practice and Policy, 12*(18), 7742.

The Valley Reporter (2022). *River Watch – Measuring pH – Then and Now*. The Valley Reporter. https://www.valleyreporter.com/index.php/news/local-news/17174-river-watch-measuring-ph-then-and-now

Tilbury, D. (1993). Environmental Education: Developing a Model for Initial Teacher Education *(Doctoral Dissertation)*. United Kingdom: University of Cambridge.

Tilbury, D., Coleman, V., & Garlick, D. (2005). *A National Review of Environmental Education and its Contribution to Sustainability in Australia: School Education*. Australian Research Institute in Education for Sustainability (ARIES).

Tilbury, D., & Walford, R. (1996). Grounded Theory: Defying the Dominant Paradigm in Environmental Education Research. In M. Williams (Ed.), *Understanding Geographic and Environmental Education: The Role of Research* (pp. 51–64). Cassels.

Tristão, V. T. V., & Tristão, J. A. (2016). The Contribution of NGOs in Environmental Education An Evaluation of Stakeholders Perceptions. *Ambiente & Sociedade, 19*, 47–66.

Vulliamy, G. (1987). Environmental Education in Third World Schools: Rhetoric or Realism?. *Environmentalist, 7*(1), 11–19.

Wals, A. E., Beringer, A., & Stapp, W. B. (1990). Education in Action: A Community Problem-Solving Program for Schools. *The Journal of Environmental Education, 21*(4), 13–19.

WCED (1987). *Our Common Future. World Commission on Environment and Development*. Oxford University Press.

Wilke, R. J., Peyton, R. B., & Hungerford, H. R. (1987). *Strategies for the Training of Teachers in Environmental Education* (No. 25). UNESCO.

4 Education and the Environment in the 1990s

Rethink and Engage

Introduction: The Emergence of Education for Sustainable Development

The 1990s saw the end of the Cold War caused by the dissolution of the Soviet Union, and the widespread proliferation of communication channels such as the Internet. In this period of relative peace and prosperity, new environmental issues began to catch the attention of the public, such as protecting tropical rainforests from destruction, biodiversity conservation, as well as the major concern of global warming as an aspect of climate change. Having emerged in the 1980s, the notion of sustainability and sustainable development entered public consciousness in the 1990s. Although these concepts did not take root in educational policy or practice during the decade, debates about the influence and implications of sustainability for environmental learning dominated academic journals and professional magazines of the time.

The United Nations Conference on Environment and Development (UNCED) held in Rio in 1992, also known as the Earth Summit, redefined the issues identified at Stockholm within the new language of sustainable development. Chapter 36 of Agenda 21 (see Figure 4.1) was a key contribution of the Earth Summit, consolidating arguments that education is critical to the achievement of sustainable development and identifying core strategies to improve learning opportunities in this area (Tilbury, 2012). Agenda 21 was seminal in that it provided a basis for international collaboration as well as a case for investment in learning for change. At the time, however, Smyth (1999) recalls that many in government and NGO bodies were committed to the call for Education for Sustainable Development (ESD) but misunderstood the role and process of education, expecting it to be a linear or infusion process that determined behavioural outcomes. It could be said that policies remained dominated by a reformist agenda that was developing socially critical teeth but had not yet transitioned towards transformative goals. Notably, at Rio plus 5 in 1997, UNESCO reported that education seemed to be 'the forgotten priority of Rio' since there had been little national reporting of action or global funding to advance its development (Tilbury & Cooke, 2005).

DOI: 10.4324/9781003467007-4

Chapter 36 of Agenda 21 is a key document from the 1992 Earth Summit. The chapter focuses on promoting education, awareness, and training for sustainable development, underscoring the pivotal role of Environmental Education in fostering a sense of responsibility and understanding among individuals and communities. Its importance lies in cultivating a mindset that recognises the interdependence of social, economic, and environmental factors. By emphasising education, Chapter 36 aims to empower people with the knowledge and skills needed to make informed decisions, participate in sustainable practices, and contribute to the preservation of the planet (United Nations, 1993).

Figure 4.1 Chapter 36 in Agenda 21

Image source: United Nations, (1993)

The Great Divide

While ESD was championed by UNESCO, the 1990s saw tensions between Environmental Education and ESD emerge, from being one and the same, to being subsets of each other, with some pointing to the more radical, critical, and liberatory educational edges of this new ESD approach. Tilbury (1995), for example, interpreted sustainability as a new framework that challenged the apolitical, naturalist, and scientific learning associated with Environmental Education in the previous decade. Tilbury also argued that ESD brought a new pedagogical lens to the learning dynamic, which awakened the interest of the learners by giving them choices on how to respond to the global crisis. Others pointed to how ESD discourses emphasised North-South inequities and interdependencies (González-Gaudiano, 1998).

A first major attempt to realise some convergence between Environmental Education and ESD, as well as to cross boundaries between school-based and community-based learning in relation to the environment, occurred at the 1992 *World Congress for Education and Communication on Environment and Development* (ECO-ED). The conference took place only months after the Rio Earth Summit in Toronto (Canada) and was sponsored by UNESCO, UNEP, and the International Chamber of Commerce (ICC) (McCrea, 2006). It can be seen as the first global conference to engage education communities, NGOs, communication systems, and governments with Agenda 21 in general, and more specifically, to implement Chapter 36.

Despite these attempts to create synergies, the 1997 Tbilisi Plus 20 conference in Thessaloniki, Greece, still showed many of the divides that had arisen as a result of the sustainability perspectives entering educational frames. Some stakeholders sought to make ESD the core label and framework for advancing efforts in environmental learning. This idea gained the support of many member states and UNESCO itself which provided special funds to increase support and visibility of ESD. Others present rejected this positioning and extended the debate to the academic literature.

Bob Jickling, for example, in *'why I do not want my children to be educated for sustainability'* (1992), highlighted the need for more dialogue and discussion on the role and purpose of education. For others, such as Huckle (1991), ESD had generated irreconcilable interpretations of education, with technocentrists on the one side, and ecocentrists on the other. Paraphrasing David Orr, this difference was between a tendency towards a global technocracy who wanted a more efficient path of development, and those who wanted a regeneration of civic culture alongside the rise of an ecologically literate and competent citizenry who could understand global issues (Orr, 1992).

Underlying the debates of the time is what Disinger (1990) and Jickling (1992) described as the potential oxymoron between 'sustainable' and 'development', highlighting that the underlying assumption of the need for economic growth is not sustainable on a finite planet. Initially highlighted by academic scholars and commentators, this tension that plagued Environmental Education efforts in the 1970s and 1980s, became more visible to policymakers and educationists during the 1990s.

A key publication at the end of this decade came from the IUCN which in 1999 organised an 'ESDebate', one of the first international online debates in the field (Hesselink et al., 2000). This novel publication included summaries of the main issues arising from the online discussion on the changing perspectives on Environmental Education in the light of the global initiative on sustainable development. Fifty experts from around the world discussed in five rounds how they saw the field evolving. The book had an accompanying CD-ROM allowing readers to delve into the points made by each participant. It also contained a top-10 list of books and sites on the subject. While ESD was becoming a major force in international policy circles, it had little traction in educational practice at the time compared to Environmental Education. Whereas the concept of Environmental Education resonated and had some meaning in educational practice and among the general public, in many countries, ESD did not. It is telling that while ESD was emerging, organisations like the North American Association for Environmental Education (NAAEE) did not change their name and held on to the Environmental Education label.

Critical Pedagogical Shifts

Important for moving the social dimension of education dialogues forward was the work of the Brazilian educator and philosopher Paulo Freire, who brought a social liberating approach to education through the concept of 'conscientisation'. This approach voiced an awareness of communities living in poverty, and the marginalisation of illiterate populations who lack the power and agency to improve their livelihoods (Freire, 1970). While Freire's work in questioning the power relationships in the education process itself began to emerge in the 1970s (see Chapter 2), it was not until the 1990s that critical pedagogical debates began to gain traction in environmental learning and education and become more closely aligned to ESD theoretical frameworks. Such pedagogies emphasise the importance of *how* the learner is engaged in the learning process and problematise the traditional power of the teacher over the learner. Helped by the development of new teaching resources, learner-centred and cross-curriculum-driven approaches became visible in school, college, and university practice, though they faced resistance to established educational structures, such as tight national curricula frameworks and overloaded assessment requirements, making it difficult for teachers to innovate the learning experience (see Figure 4.2).

'Reaching out' was the World Wide Fund for Nature (WWF) UK's program for professional development based on resources developed in the early 1990s. It provided a critical education for sustainability, seeking to engage teachers with ethical, philosophical and pedagogical foundations of education for sustainability. With a cross-curriculum approach, the program had the aim of integrating sustainability principles into all education and training environments, especially through linking education with political projects beyond the school. Although widely celebrated by the academic community, the programme gained limited success, with teachers struggling to accept it due to the conservative educational reforms of the time (Huckle, 1998). Rather than cross-curriculum themes such as sustainability, teachers were still focussing on effective delivery and assessment of the national curriculum's core and foundational subjects. It took another ten years before this material rippled into influence into school textbooks and reframed what was considered good practice. With the help of the WWF, the texts became influential in countries such as China, Hungary, South Africa, Tanzania, and the UK with its materials widely used in initial teacher education and pre-service training.

Figure 4.2 Learner-centred, cross-curriculum, and critical education.

Source: Huckle (1993)

The decade also saw scholars like Paul Hart, Ian Robottom, Bob Jickling, Bob Stevenson, Noel and Annette Gough, and Rick Mrazek converge on the need to reframe research inquiry in Environmental Education. These scholars continued to question the scientific stronghold and the narrowly constructed empirical analytical research, instead demonstrating how alternative forms of inquiry could be legitimate and valuable. Daniella Tilbury spoke of the contradictions in the purpose and practice of research (Tilbury, 1993), whilst Karen Malone (1999) introduced the idea of Environmental Education research as a form of activism. By the late 1990s, Paul Hart and Kathy Nolan concluded in their review of Environmental Education research that participatory and inquiry-based approaches, underpinned by different ontological and epistemological assumptions, were on the rise (Hart & Nolan, 1999). The authors also noted that the literature was increasingly critical of the organisational structure of schools and teacher education systems, which were seen to be grounded in a different world and planetary views to those advanced by Environmental Education.

These socially critical movements also drew attention to feminist perspectives questioning power relationships between men and women. In the 1990s, this particularly focussed on the connection between the domination of women and the domination of nature by 'Man' with the view that harmony with nature requires critically addressing inequities in human relations. This can be seen in the history of Environmental Education, where it is important to note how the Western background and expertise of the *founders* of Environmental Education – white men with scientific backgrounds – influenced how Environmental Education evolved through certain biases and assumptions in regards to representations of universalised subjects, such as 'Man' and 'nature' (Gough, 1997; Gough & Whitehouse, 2003).

Ecofeminism, similar to the other critical education research frames being advanced in the 1990s (Gough & Robottom, 1993), focused on politics and power, paying special attention to excluded and formerly silenced groups. Researchers set out to challenge dominant approaches to learning and definitions of knowledge as well as methodologies of inquiry (Gough, 1999). Several leading female scholars began formulating ecological feminist responses to environmental impacts of encroaching modernisation. For instance, ecofeminist scholars such as Vandava Shiva highlighted the issue of reductionist, mechanistic science and the attitude of conquest over nature as an expression of capitalist patriarchy (Mies & Shiva, 1993). In a similar vein, Li (2007) highlights the necessity to critique the idea that Environmental Education is about celebrating and preserving nature. Rather, she argues, it is about highlighting the need to examine, critique, and rectify the unequal social relations embedded in contemporary society. These ideas gained traction in the 2000s as the need for a fundamental change in how we live on Earth became clearer to more people.

At the same time, place-based approaches were becoming more important, with figures such as Wendell Berry (1997) and David Orr (1992) highlighting the importance of reinhabiting place. This is also seen in critical place-based pedagogy (Gruenewald, 2003), which merged critical pedagogies and a place-based approach with an emphasis on experiencing the environment physically (Payne, 1997). These frames took almost three decades to capture the interest of mainstream practitioners.

Early Childhood Education

The 1990s also began to more seriously recognise that early childhood education and care were critical. Research documented how the early years were crucial to developing attitudes and behaviours in Environmental Education. Joy Palmer and her colleagues pointed to how young children have an innate ability to see things in a relational and moral way early on and seem to lose this ability as they spend more time in schools (see Palmer, 2002; Palmer & Neal, 1994). In the US, initiatives were undertaken to reinforce the importance of promoting a sense of wonder amongst children, with the early years seen as pivotal in nurturing relational ways of thinking, playfulness, and art in the environment (Wilson, 1996). This work led to a rise in early childhood

Environmental Education activities focused mostly on exposing children to nature and building emotional connections to it. In Australia, Julie Davis (1998) pointed to the dangers of idealising nature and seeking to inculcate values, problematising nature-oriented education in the early years. She argued that through education children are already being colonised by exploitative ideas and practices towards each other and the environment, and that much work needed to be done to stress environmental perspectives for children in curriculum theory, policy, and practices. Roy Ballantyne's research documenting the impact of children on parents in regards to Environmental Education was also seminal and pointed at the need to bridge the intergenerational divide through education (Ballantyne et al., 1998).

Teacher Education and Educational Innovation

The 1990s also saw teacher professional development enter the agenda with force – with UNESCO defining it the Priority of Priorities (UNESCO-UNEP, 1990). UNESCO Asia Pacific generated the first framework in support of teacher education for the environment, with Fien and Tilbury (1996) interpreting the inclusion of Environmental Education into teacher education as a dynamic educational innovation process. Alongside this, the Environment and School Initiatives (ENSI), founded by the OECD in 1986, had brought together leads in government agencies responsible for Environmental Education from Australia, Canada, South Korea, and several European Countries. This initiative supported schools in engaging with environmental issues as an educational innovation, rather than simply as a scientific problem to be solved (Affolter & Varga, 2018). An example of this was exploring how action research could help teachers and educators to link environmental concerns with global issues and to inject some innovative pedagogical practices in support of citizenship education (Affolter & Varga, 2018). This approach offered a contrast to efforts in the 1980s that mapped specific educational content to be addressed through teacher professional development on matters of the environment (Wilke et al., 1987). By the middle of the 1990s, UNESCO-UNEP established an International Environmental Education Programme for Teacher Education in recognition of the pivotal role teachers play in mainstreaming Environmental Education (Fien, 1995).

Japan's National Institute for Educational Research also turned its attention to teacher education in the context of Environmental Education. Over a period of three years in the mid-1980s, it convened researchers from across the Asia-Pacific in Tokyo to unpack key trends, needs, and opportunities for extending Environmental Education practice through research. Its annual publications were often referenced and used to inform national policy developments and efforts in the region.

'Muda o Mundo, Raimundo!' was an initiative of WWF Brazil, the Federal Ministry of Education and the Brazilian Institute for Environment and Renewable Resource (IBAMA). The initiative sought to embed Environmental Education in the primary school curriculum in ways that engaged the learner with real world issues. It adopted a cross-curricular approach and placed emphasis on citizenship engagement. It took pupils and teachers out of their classrooms and into their neighbourhoods, thus linking global environmental issues to local concerns. The work was guided by a teacher guide and supported by the Brazilian Federal Constitution that required Environmental Education to be taught in all primary schools. The initiative launched in 1995 with a series of regional training workshops in Brasilia, Parana, Pernambuco, Maranhao, Rio de Janeiro, Rondonia and Sao Paulo. It developed a network of trained educators who could 'ripple' the educational innovation across schools (Fien et al., 1999).

Figure 4.3 Change the World, Raimundo!

NGOs also became involved in this space and, through education partnerships, harnessed the potential of teacher networks in catalysing change across the national education offerings. In Brazil, the 1995 programme '*Muda o Mundo, Raimundo!*' proved effective in embedding environmental issues in citizenship education efforts and framing the environment as a cross-curricular concern. This initiative was ahead of its time and was one of the first to map a theory of change in support of embedding Environmental Education across primary schools and was accompanied by pathways for teacher development in this area (see Figure 4.3).

Transformative Perspectives in Education

Underpinning these educational responses were transformative perspectives to education. Rather than the traditional practice of transmission of facts, skills, and values to student, with closed learning outcomes decided on by experts, a transformative perspective sees knowledge and understanding as being co-constructed within a social context—new learning is shaped by prior knowledge and diverging cultural perspectives (O'Sullivan, 1999). Informed by socially critical perspectives, transformative education provides space for autonomy and self-determination on the part of the learner. In this sense, a function of this form of environmental learning is to encourage students to become critically aware of how they perceive the world with the intention of fostering citizen engagement and participation in decision-making processes. Jickling and Wals (2008) point out that when deprived from this space and function, Environmental Education runs the risk of facilitating 'Big Brother' sustainable development, characterised by authoritative policy statements and government directives, transmissive goals, and authoritative approaches to learning to generate an obedient population.

From a transformative perspective, education is therefore more about teaching students *how* to question and reflect in their thinking, rather than *what* to think. In this vein, the work of Jack Mezirow and Edmund O'Sullivan on transformative learning in the mid-1990s (Mezirow, 1997; O'Sullivan, 1999) portrayed learning as a process of deep, constructive, and meaningful learning that goes beyond simple knowledge acquisition and supports critical ways in which learners consciously make meaning of their lives (Taylor, 1998). This trend gave further strength to socially critical approaches that had similar perspectives on educational change. These approaches helped shift practice away from a focus on awareness raising and engaging learners in isolated environmental activities such as picking up rubbish. Instead, efforts encourage learners to develop critical and systematic thinking skills, which address the root of the problem (such as consumerist culture in this case). Important to the process of engendering a transformative perspective is the need for learners to engage in action-based reflection with an overt agenda of social change (Gadotti, 1996; Tilbury & Cooke, 2005). Also significant is the focus on values clarification (rather than values education or inculcation), whereby learners critically assess their own beliefs, values, and worldviews using forms of dialogue to explore the inevitable tensions and differences between them (Kollmuss & Agyeman, 2002).

In a similar vein, new fields began to gain traction, such as futures education that considers potential futures through the exploration of values, visions, and drivers for change (Hicks, 1998; Hicks & Holden, 1995). In line with the future-oriented emphasis of the sustainable development definition of the Brundtland report (WCED, 1987) – 'meeting the needs of the present without compromising the ability of future generations to meet their own needs' – this perspective was thought to transform the way people relate to their future, helping to motivate engagement and create opportunities for change. While this strand became a fundamental characteristic of learning for sustainability in later decades, the seeds were planted in the 1990s.

Community-based Learning and Education

Another educational response was the involvement of people outside of formal education, through learning experiences in the local communities of learners. The rise of community-based learning took place with the increasing realisation that the local and global are deeply intertwined and connected. While global issues such as inequality, injustice, conflict, and the environment are found in various parts of the world, they are also experienced locally. As the saying goes, 'think globally, act locally'. This 'glocal' perspective built upon the Tbilisi Declaration of 1977, which was the first declaration giving international recognition to the importance of community educational approaches in creating change for the environment. The Rio Declaration in 1992 and Agenda 21 further promoted the role of community education by repositioning education at the centre stage of community building for a sustainable future (Leicht et al., 2018; Pozo-Llorente et al., 2019). Governments and non-government organisations were encouraged to define their roles and establish priorities for community learning, leading to multi-stakeholder and participatory approaches that sought to improve local environmental issues. Derived from Agenda 21, *Local* Agenda 21 sought to build upon existing local government strategies and resources to implement sustainability goals. As demonstrated by Daniella Tilbury and colleagues in the context of Australia, Local Agenda 21 had an important impact on how local communities engaged with sustainability issues, encouraging municipalities to participate, influence, and share the decision-making process (2005). Indicative of the importance being placed on local environmental issues, Bob Evans and Julian Agyeman launched the journal Local Environment in 1996, which had as its lead article 'Local agenda 21, compulsory competitive tendering and local environmental practices' (Patterson & Theobald, 1996).

An important aspect of community education is engaging the community in participatory learning, which aims to build capacity for change towards sustainability. William Stapp and colleagues developed an action learning model for community problem solving, which they initiated in the late eighties, engaging young people in a participatory initiative to resolve a socio-environmental problem perceived in their own community (1996). Beyond problem-solving, an action learning approach requires constant reflection and an action-oriented focus. The added 'participatory' component informed by the Participatory Action Research (see Fals Borda, 1988) emphasises the importance of engaging with political action aimed at enacting participants' agency to bringing about radical changes in asymmetrical power relations and narratives that maintain oppressive and exploitative conditions.

The decade of the 1990s also saw those engaged in conservation acknowledging the significance and role of education in attaining biodiversity goals. This was a key milestone for those campaigning to profile learning and communication as a key conservation measure. Conservation professionals may have struggled with embracing the new socially critical approaches to education but they brought with them strategic mindsets that influenced how education and learning experiences were framed and advanced as a policy agenda. For example, the IUCN convened an international group of experts through its Commission for Education and Communication, which generated strategic plans for learning in biodiversity as well as sought to embed education and communication across biodiversity strategies and commitments (see Figure 4.4). This

The IUCN published a key strategic document called Environmental Education and Sri Lanka, that positioned education as a key social driver for sustainable development that sought to engage professionals and people for all
walks of life (IUCN, 1998). The document identified a five-year framework
for how to engage policymakers, media, teachers, students, law enforce-
ment officers, professionals, NGOs, rural communities, and the corporate sector in conservation through education. The document formalised
the role of education and learning as a key strategy for the attainment of
conservation.

Extracts from the text reveal how education and communication were to
be 'used to save threatened species and ecosystems' and 'spread the message
of conserving the natural environment' (IUCN, 1998, p. 1). The focus was
on 'instilling knowledge, skills and values', 'enhancing knowledge of biological
diversity and the need to conserve it', 'increasing awareness and concern
among the public about natural landscapes and resources', and 'strengthening
information output on species and ecosystems'. These extracts show that
socially critical or transformative education practices had not yet filtered into
mainstream conservation education policy or practice.

Figure 4.4 Conservation and Education.

Source: IUCN (1998)

served as an indicator of how education and learning were formally recognised
by natural environment professionals as a key measure or deliverable towards
protecting a biodiverse world.

Friends of the Earth, the world's largest grassroots environmental network
at the time, convened China's first International Conference in Environmental Education in 1993. The event which was supported by the Ministry for
Basic Education brought together leading figures in the field from across the
continents as well as doctoral students researching environmental learning
in schools, national parks, and teacher education programmes across China.
It was not long after this seminal event that WWF China offered a training
programme for teacher educators based at Chinese Normal Universities. The
courses led by John Huckle and co-facilitated with others who attended the
1993 Conference, interestingly brought critical reflective lenses, and supported pedagogical innovation across the education system. Independent
evaluations recorded the positive impact of this programme and credited it
with advancing Environmental Education in schools in key Chinese cities
(Fien et al., 1999).

Also worthy of mention is the evaluation of the Worldwide Fund for Nature's (WWF) global educational programmes. The evaluation sought to collate
evidence to clarify and document the contribution that Environmental Education was making to the achievement of conservation goals (Fien et al., 2001).

This was the first international, strategic, and comprehensive evaluation undertaken to qualitatively assess the impact as well as lessons learnt from the investment in Environmental Education programmes. As the evaluators explain, the findings offered insights that go beyond its immediate value to WWF, thus providing tangible evidence of impact as well as a vocabulary for evaluating Environmental Education programmes.

Moving from the 1990s into the 2000s

By the end of the decade, the understanding of teaching and learning for sustainable development had moved beyond initial linear and limited conceptions, but there were still those, often with much influence and power, who held on to a 'business as usual' agenda. The terminology debates prompted by 'sustainable development' distracted from the real issues at hand; whether education would accommodate new thematic learning and specialist courses into its offerings, or whether it would reframe learning relationships and educational experiences so that all pupils could be engaged in learning for a more sustainable planet.

One aspect had become increasingly clear; however, while the last two decades had led to a more environmentally aware population, people still lacked the necessary knowledge and skills to know which action to take (Gigliotti, 1990). There was also little evidence to show that particular values corresponded with specific actions, so the decade also saw a move away from more instrumentalist objectives of acquiring a specific set of values and attitudes for the environment. Instead, the 1990s saw the theme '*rethink and engage*' dominate practice and call for more interpretive and critical lines of inquiry through environmental learning and education (A. Gough, 1999; Hart & Nolan, 1999; Palmer, 2002). This meant greater support for an educational approach which not only considered immediate environmental improvement as an actual goal but also started to consider educating for the long term. The result was a very different pedagogical style and learner focus which critiqued the way we see the environment, the way we see one another, and the way society engages with the natural world. See Figure 4.5 for a summary of education and the environment in the 1990s.

The turn of the century into the 2000s saw perhaps the most significant changes to educational frames and responses to environmental issues. As the following chapter will show, the decade consolidated and mainstreamed many of the more emergent approaches and marginal narratives that had been brewing over the previous twenty years. Issue-resolution learning, single action outcomes, or behaviour change approaches still existed but were no longer the dominant aspirational goals for the Environmental Education movement, instead more integrative and emancipatory approaches were on the rise.

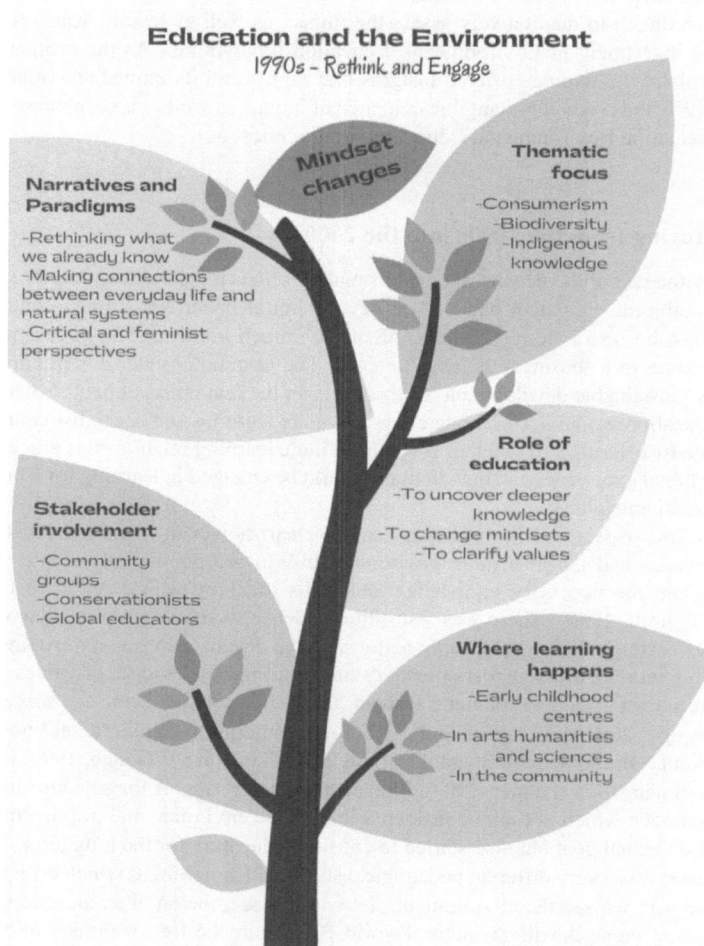

Figure 4.5 Summary of education and the environment in the 1990s

References

Affolter, C., & Varga, A. (2018). *Environment and School Initiatives: Lessons from the ENSI Network-Past, Present and Future*. Environment and School Initiatives-ENSI.

Ballantyne, R., Connell, S., & Fien, J. (1998). Students as Catalysts of Environmental Change: A Framework for Researching Intergenerational Influence Through Environmental Education. *Environmental Education Research, 4*(3), 285–298.

Berry, W. (1997). *The Unsettling of America: Agriculture and Culture.* Sierra Club Books.

Fals Borda, O. (1988). *Knowledge and Peoples Power.* New Horizons.

Davis, J. (1998). Young Children, Environmental Education and the Future. *Early Childhood Education Journal,* 26, 117–123.

Disinger, J. F. (1990). Environmental Education for Sustainable Development? *The Journal of Environmental Education, 21*(4), 3–6.

Fien, J. (1995). Teaching for a Sustainable World: the Environmental and Development Education Project for Teacher Education. *Environmental Education Research, 1*(1), 21–33.

Fien, J., Scott, W., & Tilbury, D. (1999). *Education and Conservation: An Evaluation of the Contributions of Educational Programmes to Conservation Within the WWF Network.* WWF.

Fien, J., Scott, W., & Tilbury, D. (2001). Education and Conservation: Lessons from an Evaluation. *Environmental Education Research, 7*(4), 379–395.

Fien, J., & Tilbury, D. (1996). *Learning for a Sustainable Environment: An Agenda for Teacher Education in Asia and the Pacific.* academia.edu. https://www.academia.edu/download/31721832/105607E.pdf

Freire, P. (1970). *Pedagogy of the Oppressed.* Continuum.

Gadotti, M. (1996). *Pedagogy of Praxis: A Dialectical Philosophy of Education.* State University of New York Press.

Gigliotti, L. M. (1990). Environmental Education: What Went Wrong? What Can Be Done? *The Journal of Environmental Education, 22*(1), 9–12.

González-Gaudiano, E. (1998) *Centro y periferia de la educación ambiental: un enfoque antiesencialista.* Mundi-Prensa.

Gough, A. (1997). Founders of Environmental Education: Narratives of the Australian Environmental Education Movement. *Environmental Education Research, 3*(1), 43–57.

Gough, A. (1999). Recognising Women in Environmental Education Pedagogy and Research: Toward an Ecofeminist Poststructuralist Perspective. *Environmental Education Research, 5*(2), 143–161.

Gough, A. G., & Robottom, I. (1993). Towards a Socially Critical Environmental Education: Water Quality Studies in a Coastal School. *Journal of Curriculum Studies, 25*(4), 301–316.

Gough, A., & Whitehouse, H. (2003). The "Nature" of Environmental Education Research From a Feminist Poststructuralist Viewpoint. *Canadian Journal of Environmental Education (CJEE), 8*(1), 31–43.

Gruenewald, D. a. (2003). The Best of Both Worlds: A Critical Pedagogy of Place. *Educational Researcher, 32*(4), 3–12.

Hart, P., & Nolan, K. (1999). A Critical Analysis of Research in Environmental Education. *Studies in Science Education, 34*(1), 1–69.

Hesselink, F., van Kempen, P. P., & Wals, A. (Eds.). (2000). *ESDebate: International Debate on Education for Sustainable Development.* IUCN.

Hicks, D. (1998). Postmodern Education: A Futures Perspective. *The American Behavioral Scientist, 42*(3), 514–521.

Hicks, D., & Holden, C. (1995). *Visions of the Future: Why We Need to Teach for Tomorrow.* Trentham Books Limited.

Huckle, J. (1991). Education for Sustainability: Assessing Pathways to the Future. *Australian Journal of Environmental Education, 7*, 43–62.

Huckle, J. (1993). *Our Consumer Society (What We Consume)*. WWF/Richmond Publishing Company.

Huckle, J. (1998). *The WWF-UK Reaching Out Programme. Futures Education, World Yearbook of Education*.

IUCN. (1998). *Education and Communication to Conserve Biodiversity in Sri Lanka*. IUCN.

Jickling, B. (1992). Viewpoint: Why I Don't Want My Children to Be Educated for Sustainable Development. *The Journal of Environmental Education, 23*(4), 5–8.

Jickling, B., & Wals, A. E. J. (2008). Globalization and Environmental Education: Looking Beyond Sustainable Development. *Journal of Curriculum Studies, 40*(1), 1–21.

Kollmuss, A., & Agyeman, J. (2002). Mind the Gap: Why Do People Act Environmentally and What Are the Barriers to Pro-Environmental Behavior? *Environmental Education Research, 8*(3), 239–260.

Leicht, A., Combes, B., Byun, W. J., & Agbedahin, A. V. (2018). From Agenda 21 to Target 4.7: The Development of Education for Sustainable Development. In A. Leicht, J. Heiss & W. J. Byun (Eds.), *The Development of Education for Sustainable Development. Issues and Trends in Education for Sustainable Development*. UNESCO.

Li, H. L. (2007). Ecofeminism as a Pedagogical Project: Women, Nature, and Education. *Educational Theory, 57*(3), 351–368.

Malone, K. (1999). Environmental Education Researchers as Environmental Activists. *Environmental Education Research, 5*(2), 163–177.

McCrea, E. (2006). *The Roots of Environmental Education: How the Past Supports the Future*. Environmental Education and Training Partnership. University of Wisconsin-Stevens Point.

Mezirow, J. (1997). Transformative Learning: Theory to Practice. *New Directions for Adult and Continuing Education, 1997*(74), 5–12.

Mies, M., & Shiva, V. (1993). *Ecofeminism*. Zed Books.

O'Sullivan, E. (1999). *Transformative Learning: Educational Vision for the 21st Century*. philpapers.org.

Orr, D. W. (1992). *Ecological Literacy: Education and the Transition to a Postmodern World*. State University of New York Press.

Palmer, J. (2002). *Environmental Education in the 21st Century: Theory, Practice, Progress and Promise*. Routledge.

Palmer, J., & Neal, P. (1994). *The Handbook of Environmental Education*. Routledge.

Patterson, A., & Theobald, K. S. (1996). Local Agenda 21, Compulsory Competitive Tendering and Local Environmental Practices. *Local Environment, 1*(1), 7–19.

Payne, P. (1997). Embodiment and Environmental Education. *Environmental Education Research, 3*(2), 133–153.

Pozo-Llorente, M. T., Gutiérrez-Pérez, J., & de Poza-Vilches, M. F. (2019). Local Agenda 21 and Sustainable Development. In W. Leal Filho (Ed.), *Encyclopedia of Sustainability in Higher Education* (pp. 1126–1135). Springer International Publishing.

Smyth, J. (1999). Is There a Future for Education Consistent with Agenda 21? *Canadian Journal of Environmental Education, 4*(1), 69–82.

Stapp, W. B., Wals, A. E. J., & Stankorb, S. L. (1996). *Environmental Education for Empowerment: Action Research and Community Problem Solving.* Kendall and Hunt Publishing.

Taylor, E. W. (1998). *The Theory and Practice of Transformative Learning: A Critical Review.* ERIC, 374.

Tilbury, D. (1993). Contradictions in the Purpose and Practice of Environmental Education Research. *International Conference in Environmental Education.*

Tilbury, D. (1995). Environmental Education for Sustainability: Defining the New Focus of Environmental Education in the 1990s. *Environmental Education Research,* 1(2), 195–212.

Tilbury, D. (2012). *Higher Education for Sustainability: A Global Overview of Commitment and Progress.* In GUNI (Ed.), *Higher Education in the World* (pp. 18–22). Palgrave Macmillan.

Tilbury, D., Coleman, V., Jones, A., & MacMaster, K. (2005). *A National Review of Environmental Education and Its Contribution to Sustainability in Australia: Community Education.* Australian Research Institute in Education for Sustainability (ARIES).

Tilbury, D., & Cooke, K. (2005). *A National Review of Environmental Education and its Contribution to Sustainability in Australia: Frameworks for Sustainability – Key Findings.* Australian Government Department of the Environment and Heritage and Australian Research Institute in Education for Sustainability (ARIES).

UNESCO-UNEP (1990). *Environmentally Educated Teachers: The Priority of Priorities* (Nos. Connect, XV, 1–3). UNESCO-UNEP.

United Nations (1993). *Agenda 21: Earth Summit—The United Nations Programme of Action from Rio.* United Nations Publications.

WCED (1987). *Our Common Future. World Commission on Environment and Development* Oxford University Press.

Wilke, R. J., Peyton, R. B., & Hungerford, H. R. (1987). *Strategies for the Training of Teachers in Environmental Education* (No. 25). UNESCO.

Wilson (1996). *Starting Early: Environmental Education During the Early Childhood Years.* ERIC Dig. http://theiwrc.org/wp-content/uploads/2010/07/Volume-23-No.-2-Summer-2000.pdf#page=25

5 Education and the Environment in the 2000s
Connect and Change

Introduction: Education in a Globalising World

The 2000s saw an accelerating globalisation fuelled by exponential growth of the Internet and strong economic development in South-East Asia and China. In addition to a homogenisation of cultures through a convergence towards consumerism, globalisation brought fragmentation and polarisation as well through terrorism, epitomised by 9/11. Both trends, coupled to increased wealth inequality, contributed to increasing social and economic disparities between people and countries. The twin pillars of development and environment – already identified in Stockholm 1972, and forged into the concept of sustainable development in the 1980s and 1990s –also affected education and the environment. This became manifest in the agreement within the UN to call for a Decade of Education for Sustainable Development (DESD). At the same time, with increasing individualisation and economic and cultural uniformity, alongside the availability of a rapidly growing amount of information ('the information age'), the purpose of education, and its role in an increasingly complex world was put into question. Currents such as ecofeminist and decolonial pedagogies strengthened their foothold in educational debates, further critiquing the way we see and relate to each other and the environment, and pointing at the role of power imbalances in how these relations are shaped (Harvester & Blenkinsop, 2010; Selby, 2008).

The Increasing Influence of Education for Sustainable Development

The 2000s saw significant momentum in Education for Sustainable Development (ESD). Internationally, the seeds of ESD had been planted during the Rio Earth Summit in 1992 and Agenda 21 and found fertile soil during the World Summit for Sustainable Development (also referred to as Rio Plus 10), which took place in Johannesburg in 2002. As noted by Tilbury and Cooke (2005), summit discussions reflected how education in the context of sustainability had evolved from mostly reorienting curriculum and training, towards

DOI: 10.4324/9781003467007-5

capacity building and learning-based strategies for change. In other words, it was no longer just about developing sustainability literacy or receiving qualifications in this area, but about understanding education as an approach to advance social change. The latter involves processes that question mindsets and untenable relationships with our planet and look at ways in which we can bring about transformative socio-economic change.

The DESD (2005–2014) was announced at the Johannesburg summit and UNESCO became a main driver of ESD as the lead agency executing the DESD. It organised numerous international events, installed UNESCO Chairs across a number of specialist disciplines, and facilitated an extensive monitoring and evaluation program of policies and practices (for an overview of some key evaluation reports see UNESCO (2021b)). At the time, Stephen Sterling was quick to point out that sustainability cannot simply be 'added-on' to disciplinary concerns or even be 'built-in' to existing structures and curricula. Rather, sustainability requires a fundamental rethinking and whole-system redesign of education, schooling, teaching, and learning, and the way institutions and the people in them behave (see Figure 5.1). Sustainability also calls for new competences relating to values, systems thinking, critical reflective practice, and engagement in social change (Sterling, 2004).

It should be acknowledged that while UNESCO has consistently advocated for ESD since it was first coined, Education *for Sustainable Development, Education for Sustainability,* and *Sustainability Education,* have often been used interchangeably, with some interpretations placing more emphasis on critical approaches to learning for environment and sustainability. Increasingly, questions in education became 'what is the role of education in creating changes socially' and 'what pedagogies can critique the way we see and relate to the environment, the natural world, and one another'. Answering such questions led to a reliance on newer educational methodologies at the heart of educational

Based on work by Sterling (2005), Linking Thinking is a professional development toolkit that recognises the complexity of sustainability issues, and that these cannot be easily 'solved'. Instead, in an interdependent world, what is required is learning to think in a more connected way (Dornan et al., 2009). The toolkit promotes 'joined up thinking' skills in problem solving, based on real-world issues such as climate change, food, and resource use that are relevant to pupil's lives, as well as encouraging connections across the curriculum. Linking Thinking is one of the first toolkits to lay the groundwork for systems thinking in learning for sustainability.

Figure 5.1 Connectivity, real world thinking, in complexity.

Source: Sterling (2005)

experiences, including systems thinking, values clarification, and critical reflective practice, which sought to challenge power structures and placing praxis, and the practical application of this learning (Tilbury & Wortman, 2005).

In response to concerns surrounding globalisation and unemployment, Lifelong learning (LLL) became popular in the 2000s (International Labour Organisation, 2019). The LLL concept recognises that education and training must be flexible and oriented towards a lifetime of learning, rather than a single career job. The release of the influential Delors report (Delors, 1996) brought LLL into policy discussions. Interestingly, similar critiques and tensions to ESD can be seen in LLL, with critiques as to the concept being imposed on the Global South by the West (Torres, 2004), as well as a neoliberal attempt by the state to shift the educational responsibility from itself to the individual (Orlovic Lovren & Popovic, 2018). It was towards the middle of the decade that UNESCO's Institute for Lifelong Learning (UIL) also began to orient its activities and resources for policy-makers and practitioners towards sustainable development.

People as 'Agents of Change'

Closely tied to the notion of *Education for Sustainability* was a move away from seeing people as 'the problem' of environmental concerns that needed to be fixed, or behaviours to be changed. Instead, there was a shift towards seeing people as 'agents of change' (Jensen & Schnack, 1997; Tilbury & Cooke, 2005). This worldview not only began to influence conservation approaches (see Figure 5.2) but also invited pedagogical changes in schools by encouraging the learner's active engagement in decision-making and developing policy (Tilbury & Cooke, 2005), whilst recognising the historical and material systems that individuals (and societies) are unevenly locked into (Spaargaren, 2011). An example of this is the Action Competence approach, presented by Jensen and Schnack (1997), which is a response to an Environmental Education paradigm characterised by individualisation and behaviour modification. This approach highlights a readiness to act in a way that meets the challenges of a given situation through the development of competencies (understandings and skills) and creating learning environments that enable learners to take critical action.

A further response to the limits of traditional, science-oriented approaches to Environmental Education were initiatives that saw humans as essentially social. The field of social learning gained traction during this decade, encouraging learning that takes place in a context of divergent interests, norms, values, and constructions of reality (Wals, 2007b). This learning was seen as especially important because it appreciates that interpretations of social organisation and economic development are inevitably value-laden and aimed at achieving particular ends and serving some interests more than others. The idea is that facilitated social learning can develop knowledge, values, and action competence which

In the 2000s, learning and education continued to be recognised by conservationists as key to advancing the environment and sustainable development commitments; the focus, this time, was on capacity building.

A report by WWF on projects in Africa documented how conservation initiatives were underpinned by education and capacity-building efforts that helped communities understand the value of the natural resources around them (WWF, 2002). For example, in a project on village ecotourism in Northern Zambia, local communities were encouraged to take ownership of their own learning processes through capacity building. The focus was on empowering villagers to take part in problem-solving and informed decision-making in issues around poaching, tourism, and wildlife conservation. Reports showed that incidents of poaching decreased since local people started participating in the education and capacity-building programmes offered by the WWF-Zambia Education Project. This highlights a move in conservation to see people as enablers and change agents, rather than the problem itself.

Figure 5.2 Education and Capacity Building in Africa.

Source: Tendai Chirwa (WWF, 2002)

increase effective participation in environmental scenarios. In this way, social learning evokes critical responses (Fien & Tilbury, 2002), as well as a reflexive approach (Wals, 2007b) to help understand the invisible threads that connect socio-economic activities to the natural environment. During the 2000s, social learning gathered momentum as a process that could bring people with different perspectives and even conflicting values together to map common futures.

Faith-based organisations also entered the frame at this time as they wrestled with the ethics and responsibilities associated with environmental issues. Dimity Podger's work with the American Bahá'í Community showed the efforts of faith organisations to build more sustainable societies and how these endeavours can shape learning discourses in Environmental Education (2009). This is done through engaging with the values dimension of learning for sustainability and has influenced international agendas and initiatives underpinned by faith education. Dimity Podger's research (2009) was seminal in that it drew attention to this work and sought to define the interface between sustainability and spirituality. In a similar vein, later work by the Kenyan Organisation for Environmental Education, together with the Alliance for Religions and Conservation, launched a teachers toolkit in support of faith-based ESD unpacking how Christianity, Hinduism, and Islam inform our responsibilities in the conservation of biodiversity and natural resources (Arc, 2011). In a related manner, the cultural arm of UNESCO explored how cultural diversity and intercultural dialogue could promote learning for environmental sustainability; a framework was developed to support participatory learning that connected the intergovernmental priorities of peace, diversity, and planetary health (Mulà & Tilbury, 2009).

The theme of 'engaging people' firmly rooted itself in community, conservation, and creative settings during this decade. In parallel, corporate social responsibility agendas began to enter business education courses, with increasing attention to exploring consumer expectations and choices, reputational indexes, and reporting related to the environment (Hunting et al., 2006). It was also in the 2000s that the concept of institutions for sustainability first emerged. Stephen Dovers (2004) argued that the collective efforts of people mediated through institutions held the key to ecological sustainable societies and that without institutional change it would not be possible to move purposefully towards sustainability. This idea would take at least another 15 years to embed itself in sustainability and learning for sustainability discourses, such as in whole-of-institution approaches (Holst, 2023).

Futures Perspectives: A See-Change

The language of transformative futures began to enter the Environmental Education discourse as it was recognised that what was needed was more than just a shift of perception of relationship with nature. Instead, the focus was on 'seeing change' and creating opportunities for the learner to envision alternative futures and visualise change for sustainability. These approaches have their roots in the futures movement of the 1980s and 1990s with the work of David Hicks (see Hicks, 1998).

In 2004, New Zealand's Parliamentary Commissioner for the Environment released a think piece entitled '*See Change: learning and education for sustainability*', recognising that the future perspective in learning could empower individuals and groups and support communities to make changes for sustainability (PCE, 2004). It acknowledged that this education can be 'uncomfortable' and that there will be strong resistance in some communities. However, it argued that education for sustainability can have a transformative effect on the way we live our lives, and pointed to measures that have seen the decline of drink-driving or smoking in New Zealand. The document questioned whether there was a robust foundation to justify education for sustainability and identify the differences and tensions between Environmental Education and this area of learning as well as future directions for New Zealand. The report refers to turning the tide of deeply held beliefs and assumptions that are leading us to unsustainable paths and puts its faith on education as a driving force for a more sustainable future. It was a timely and visionary text as it recognised the power of education to shake society to the core and was critically influential given the Commissioner's remit to drive change across government policy. The document helped to reposition education for sustainability as a learning process relevant to government agencies, civil society, and local communities and breaching the boundaries of the school in the quest for sustainable living (Blewitt & Tilbury, 2013).

Challenging Dominant Discourses: Decolonisation

The fundamental critique of how individuals and societies relate to the environment can also be seen in decolonial education discourse, as a response to Western-imposed paradigms of development (de Sousa Santos, 2007). Decolonial currents address the narrow rationalities characterised by colonial and imperialist thinking (see Dussel, 1998) and specifically refer to a historical process whereby countries that were colonised by foreign powers obtain their independence. It is important to note that while countries may have been politically decolonised, neo-colonialism in education refers to contexts whereby Western paradigms have and continue to shape and influence educational systems through the process of globalisation (e.g. through colonial languages such as English and French (Obondo, 2007). This can also be seen in the perceived pressure to modernise and reform education so as to attain high international standards (Nguyen et al., 2009; Wals et al., 2022)

Aiming for epistemic plurality or multiple ways of knowing (Andreotti et al., 2011), decolonial pedagogies promote marginalised forms of knowledge, such as indigenous and local knowledge. These pedagogies have a strong tradition in Latin America in line with Freirean emancipatory pedagogies and Environmental Education (Walsh, 2010), as well as African movements such as Ubuntu (Chilisa, 2017; Le Grange, 2016; Tavernaro-Haidarian, 2019). Interestingly, decolonisation has been put forward as a future frame for environmental and sustainability education (Lotz-Sisitka, 2017), opening up opportunities for emergent, generative models for education. However, as highlighted by Lotz-Sisitka and others, a danger in decolonising Western models is replacing a dominant paradigm with a marginal paradigm, resulting in an equally homogenising and static model (Lotz-Sisitka, 2017; Lotz-Sisitka et al., 2022; Macintyre et al., 2020). Rather, it is argued that all knowledge traditions are in some way embedded with power and inequalities and are constantly changing through an exchange of ideas and practices. We can understand these diverse and dynamic expressions as an ecology of knowledge (de Sousa Santos, 2014).

While we can see openings towards decolonisation, and other ways of 'being' in the world in UNESCO documents (see the Berlin declaration UNESCO, 2021a), Silova et al. (2020) have noted that UNESCO itself represents a form of 'Re-westernisation' through the reaffirmation of the liberal western model of the universal (see Mignolo, 2013). In the same way, colonisation has left footprints of the West in the formal and higher education curriculum but also in the way that teachers are trained, thus shaping the teaching, and learning dynamic and the power relationships that underpin education. In terms of higher education, for example, Alvares and Faruqi (2014) highlight the need to rethink whose knowledge underpins curriculum, restructuring to consider who leads education institutions and reframing community relationships to reconsider who is involved.

The United Nations Decade of Education for Sustainable Development

The 2002 Johannesburg Rio Plus 10 Summit was a significant milestone from an educational perspective. The summit reviewed progress made towards sustainable development over the past 10 years and sought to work towards commitments to action (United Nations, 2000). It saw the largest ever gathering of world leaders and over 21,000 participants from 191 government, intergovernmental and non-government organisations, the private sector, academia, and the scientific community. The mere presence of these stakeholders, willing to engage in the process, demonstrated the continued interest and relevance of sustainable development (Tilbury, 2003).

Non-governmental stakeholders attending this Summit powered on the agenda of education and people engagement shaping negotiations and calling for significant investment as well as international collaboration in these areas. Tilbury and Wortman (2004) document the dialogues and the ambition for education to go beyond technical responses to promote vision, values, and participation for change (see Figure 5.3). The Johannesburg Declaration (UNESCO, 2002b) enshrined education as a foundation of sustainable development, and the Johannesburg Summit itself was considered a move towards understanding the achievement of sustainable development as a learning process (UNESCO, 2002a). Essentially, it reinvigorated global commitments and actions towards more sustainable forms of development and recognised that no nation had yet attained sustainability. At the same time, there were critical voices who lamented the lack of consensus on specific targets (Von Frantzius, 2004), and failure of the Summit to ratify the more holistic, non-anthropocentric educational framework as outlined in the Earth Charter (Kahn, 2008).

The International Union for Conservation of Nature (IUCN) Commission on Education and Communication (CEC) convened an international meeting with participants attending the Johannesburg Summit in 2002. The three-day event sought to clarify the influence of sustainability on educational practices for the environment and define what. The meeting gathered over 300 policymakers, educators, and education decision-makers from over 52 countries. It reflected the global interest in alternative forms of learning for the environment but also captured how the education community had become increasingly vocal at attracting the support of environmental decision-makers. The dialogue brought new energy to international dynamics (Paden, 2002), paving the way for the UN Decade in Education for Sustainable Development as is captured in the text 'Engaging People in Sustainability' (Tilbury & Wortman, 2005).

Figure 5.3 The influence of sustainability on educational approaches to environment.

Source: Tilbury and Wortman (2005)

The Johannesburg Declaration led to the adoption of the United Nations Decade of Education (DESD) that took place between 2005 and 2014. As a platform, the DESD aimed to embed sustainable development into all learning spheres through a reorientation of education and developing initiatives that could showcase the special role of ESD (Wals, 2009, 2012). Despite notable critiques (see González-Gaudiano, 2005; Jickling, 2006; Pérez & Llorente, 2005; Sauvé & Berryman, 2005), there was support and high expectations across many stakeholders from across the globe for the opportunities presented by the DESD. One example was the hope that it would connect adjectival educations, other than Environmental Education, to sustainable development (Mulà & Tilbury, 2009).

Whilst the professional journals, education newsletters and scholarly literature grappled with what the DESD might mean for education, critiques of practice continued. At this point, Stevenson (2007) highlighted the remaining tensions between Environmental Education and schooling systems – concerns which were later deepened by Sterling and Huckle's work calling for transforming education systems towards sustainability (Sterling & Huckle, 2014). These issues were mirrored in reviews of policies and practices of sustainability in higher education – an agenda that started to dominate academic and stakeholder policy dialogues (Ryan et al., 2010; Tilbury, 2004, p. 2012; Tilbury et al., 2005). It would take another decade for these conversations to have an impact on institutions and programmes.

Notably, the 2000s saw the DESD catalysing the development of national strategies and coordinating bodies in support of this area of learning (see Figure 5.4). Germany, for example, launched its National Plan of Action for the Decade in January 2005. The launch event was organised by the Federal Ministry of Education and Research, the German Commission for UNESCO, and the public broadcaster ZDF. Germany's Plan for the DESD was supplemented by a 'Catalogue of Measures' that listed over 60 concrete and measurable strategic activities by major stakeholders that intended to contribute to reorienting education towards sustainable development (UNESCO, 2007). Throughout 2005, the DESD national coordinating body certified over 170 ESD initiatives identified as best-practice and recognised as formal contributions to the DESD. The aim of this initiative was to promote this practice through the media coverage to raise the visibility of ESD and encourage stakeholders throughout Germany to support this commitment. Both Japan and Germany championed the UN Decade with Bonn hosting the first global intergovernmental meeting in ESD in 2009, and Tokyo the final meeting in 2014.

In 2003, just prior to the launch of the DESD, the Fifth Ministerial 'Environment for Europe' conference took place in Kiev, giving the green light for a regional strategy on ESD. Less than two years later, Ministers and other officials from education and environment Ministries from across the United Nations Economic Commission for Europe (UNECE) region adopted the UNECE strategy for ESD at their joint high-level meeting in Vilnius.

In 2006, 44 countries had established a national coordinating body for ESD. By the close of 2008, this number had increased to at least 78 countries, marking significant progress within a relatively short timeframe. Regional differences existed, with Europe and North America having a greater number of national platforms. Members of coordination bodies were typically comprised of government officials, NGOs, and educational stakeholders (such as policymakers, administrators, and sometimes teachers). Some countries also included the private sector in these structures and groupings. The extent of government involvement and the scope of ESD coordination vary, ranging from ministry oversight to decentralised regional responsibilities. National coordinating bodies operated differently, with some focusing narrowly on formal primary and secondary education, while others adopted a broader approach encompassing non-formal learning and teacher professional development across all educational levels. However, representation of labour unions, religious groups, and the mass media was generally limited within these bodies (Wals & Kieft, 2010). In some countries, for example Germany, the national coordinating bodies remained after the conclusion of the DESD and evolved to bring a sharper focus on youth voices and engagement at the heart of policy-making.

Figure 5.4 ESD national coordination bodies, structures, and policies

The UNECE strategy for ESD was to encourage countries to integrate ESD into education systems from primary to tertiary education, including vocational and adult learning (UNECE, 2005). It called on 'participatory teaching and learning methods' that empowered the learners to take action for sustainable development. The objectives of this strategy included: developing policy, regulatory, and operational frameworks to support ESD; promoting sustainable development through formal, non-formal, and informal learning; developing educator competence in ESD as well as strengthening co-operation on ESD across the UNECE. Michel Ricard (2013) noted that these ambitions were admirable but posed a significant challenge to South Eastern and Eastern Europe, particularly for children living in rural areas where access to education was limited due to a lack of financial and human resources. Ricard also noted that in the initial years, member states that were less developed industrially favoured primary education, whereas developed countries focussed more on engaging communities and adults in the process.

In another part of the world, the Pacific Regional Strategy for ESD was endorsed for the Pacific Forum Education Ministers September 2006 in Nadi Fiji (Pacific Forum Education Ministers, 2006). It offered a framework for the implementation of a regional approach to ESD and the identification of coordinating actions. This strategy was recognised as a commitment made by Pacific countries in adopting the DESD and served to convene numerous activities, many culturally centred, in support of learning for a more 'prosperous and sustainable' future. The non-formal sector played a critical role in the delivery of this regional strategy.

Parallel to this work was the field of Technical and Vocational Education and Training (TVET). This field started to become engaged with sustainability as the notions of 'green skills' started emerging, whereby the world of work was seen as a critical sphere for cultivating technologies and lifestyles that foster sustainability, rather than unsustainability. In this context, certain questions began to arise: 'what kinds of skills and competences do people working in both the private and public sector need; and how can initial education and training as well as continuing professional development of staff develop these?' The mid-term UNESCO DESD monitoring report's section on TVET concluded that companies – driven mostly by economic interests and technological innovations – were beginning to reorient themselves towards the 'green economy' and related 'green skills and jobs' and that vocational schools were responding by reorienting their curricula (Wals, 2012). Alongside key research by Fien et al. (2008), and a handbook on bridging academic and Vocational Learning by Maclean et al. (2009), there were islands of innovative practice in the sector. However, it was not until 2023 that the European Commission compiled a compendium of inspiring case studies that support the development of the competences and skills needed by employees and trainees to engage with the green transition (2023).

Moving from the 2000s into the 2010s

Over 30 years on from the 1972 Stockholm conference, both Gough (2006) and Wals (2007a) noted that the DESD policy structures were very similar to the framing of the Belgrade Charter Framework from 1975 (UNESCO, 1975), and the 1977 Tbilisi Declaration (UNESCO-UNEP, 1978). This raised the question of what had really been learnt over the last decades about the role of education in addressing environmental concerns. Towards the end of the 2000s, at the halfway mark of the DESD, it is noted that despite the ambitious agenda – with a more holistic focus on social, economic, and cultural dimensions – there is a lack of deep engagement and implementation by governments to invest and support the development of educators and researchers in ESD, as well as to develop different mechanisms to evaluate these experiences and capture lessons learned (Mulà & Tilbury, 2009; Wals, 2009). See Figure 5.5 for a summary of education and the environment in the 2000s.

The next chapter will cover the decade of the 2010s, which centres on the intersection of the educational paradigms that emerged in the 1990s and the first decade of the new millennium. This period witnessed an intensified

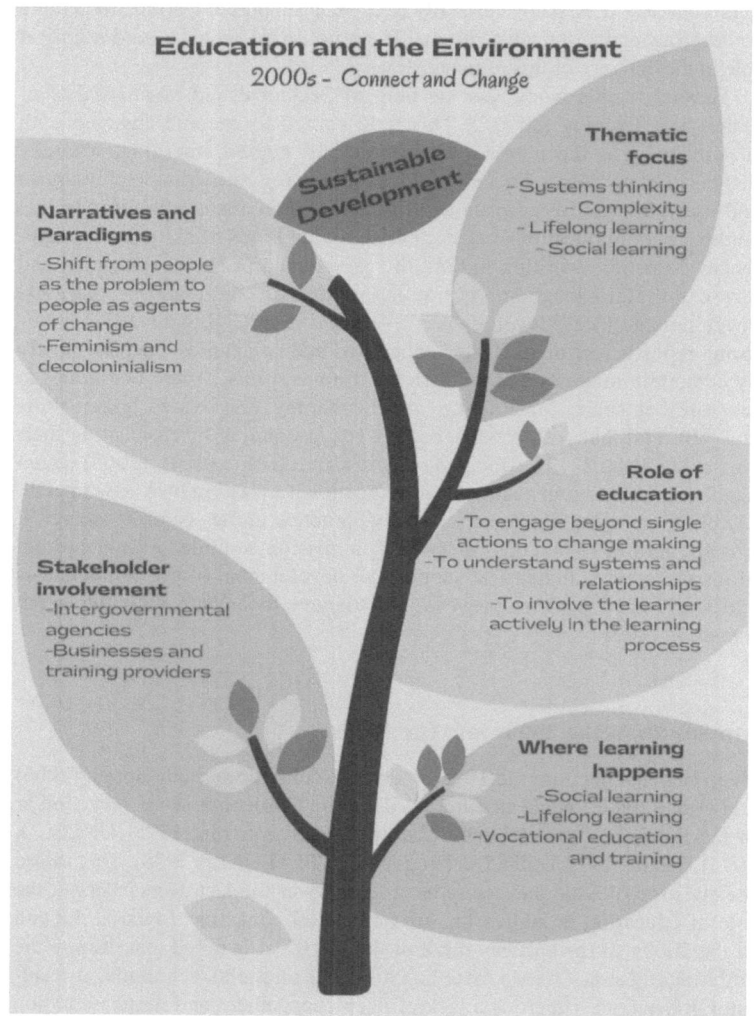

Education and the Environment
2000s - Connect and Change

Thematic focus
- Systems thinking
- Complexity
- Lifelong learning
- Social learning

Sustainable Development

Narratives and Paradigms
- Shift from people as the problem to people as agents of change
- Feminism and decoloninialism

Role of education
- To engage beyond single actions to change making
- To understand systems and relationships
- To involve the learner actively in the learning process

Stakeholder involvement
- Intergovernmental agencies
- Businesses and training providers

Where learning happens
- Social learning
- Lifelong learning
- Vocational education and training

Figure 5.5 Summary of education and the environment in the 2000s

global focus on addressing escalating sustainability concerns as a backdrop. The prominence of Sustainable Development Goals (SDGs) underscored the urgency of incorporating environmental, climatic, and social justice considerations into educational curricula. Concurrently, there was a concerted endeavour to integrate sustainability principles and practices within educational institutions, exemplified by the ascendancy of the whole-school approach. Additionally, acknowledging the constraints inherent in contemporary educational systems, voices advocating for post-colonial education, often rooted in alternative developmental frameworks, presented novel pedagogical approaches grounded in principles of relationality, plurality, and a profound ethos of respect and care for the planet. These multifaceted developments persistently scrutinised the role of education and learning in the overarching pursuit of a sustainable and global ecosystem.

References

Alvares, C., & Faruqi, S. S. (2014). *Decolonising the University: The Emerging Quest for Non-Eurocentric Paradigms*. Penerbit USM.

Andreotti, V., Ahenakew, C., & Cooper, G. (2011). Epistemological Pluralism: Ethical and Pedagogical Challenges in Higher Education. *AlterNative: An International Journal of Indigenous Peoples, 7*(1), 40–50.

Arc, K. A. (2011). *Faith-Based Education for Sustainable Development: Teachers Toolkit'Nairobi*. Kenyan Organisation for Environmental Education and Alliance for Religions and Conservation.

Blewitt, J., & Tilbury, D. (2013). *Searching for Resilience in Sustainable Development: Learning Journeys in Conservation*. Routledge.

Chilisa, B. (2017). Decolonising Transdisciplinary Research Approaches: An African Perspective for Enhancing Knowledge Integration in Sustainability Science. *Sustainability Science, 12*(5), 813–827.

de Sousa Santos, B. (2007). Beyond Abyssal Thinking: From Global Lines to Ecologies of Knowledges. *Review, 30*(1), 45–89.

de Sousa Santos, B. (2014). *Epistemologies of the South: Justice against Epistemicide*. Paradigm Publishers.

Delors, J. (1996). Learning: The Treasure Within. Report to UNESCO of the International Commission on Education for the Twenty-First-Century, Paris UNESCO 1996. *Internationales Jahrbuch Der Erwachsenenbildung, 24*(1). https://doi.org/10.7788/ijbe.1996.24.1.253

Dornan, J., Keast, J., & King, B. (2009). *Linking Thinking: Curriculum for Excellence. Joined-up Thinking for the Classroom*. WWW Scotland.

Dovers, S. (2004). *Institutions for Sustainability*. Centre for Resource and Environmental Studies, The Australian.

Dussel, E. (1998). Beyond Eurocentrism: The World-System and the Limits of Modernity. In *The Cultures of Globalization* (pp. 3–31). Ed. F. Jameson, and M. Miyoshi. Duke University Press.

European Commission (2023). *Vocational Education and Training and the Green Transition – A Compendium of Inspiring Practices*. Publications Office of the European Union.

Fien, J., Maclean, R., & Park, M.-G. (2008). *Work, Learning and Sustainable Development: Opportunities and Challenges*. Springer Science & Business Media.

Fien, J., & Tilbury, D. (2002). The Global Challenge of Sustainability. In D. Tilbury, R. Stevenson, J. Fien & D. Schreuder (Eds.), *Education and Sustainability: Responding to the Global Challenge*. IUCN.

González-Gaudiano, E. (2005). Education for Sustainable Development: Configuration and Meaning. *Policy Futures in Education, 3*(3), 243–250.

Gough, A. (2006). A Long, Winding (and Rocky) Road to Environmental Education for Sustainability in 2006. *Australian Journal of Environmental Education, 22*(1), 71–76.

Harvester, L., & Blenkinsop, H. (2010). Environmental Education and Ecofeminist Pedagogy: Bridging the Environmental and the Social. *Canadian Journal of Environmental Education (CJEE), 15*, 120–134.

Hicks, D. (1998). Postmodern Education: A Futures Perspective. *The American Behavioral Scientist, 42*(3), 514–521.

Holst, J. (2023). Towards Coherence on Sustainability in Education: A Systematic Review of Whole Institution Approaches. *Sustainability Science, 18*(2), 1015–1030.

Hunting, S., Mah, J., & Tilbury, D. (2006). *Education about and for Sustainability in Australian Business Schools: Embedding Sustainability in MBA Programs.* Australian Research Institute in Education for Sustainability (ARIES) for the Australian Government Department of the Environment, Water, Heritage and the Arts.

International Labour Organisation (2019). *Work for a Brighter Future: Report of the Global Commission on the Future of Work.* ILO.

Jensen, B. B., & Schnack, K. (1997). The Action Competence Approach in Environmental Education. *Environmental Education Research, 3*(2), 163–178.

Jickling, B. (2006). The Decade of Education for Sustainable Development: A Useful Platform? Or an Annoying Distraction? A Canadian Perspective. *Australian Journal of Environmental Education, 22*(1), 99–104.

Kahn, R. (2008). From Education for Sustainable Development to Ecopedagogy: Sustaining Capitalism or Sustaining Life? *Green Theory & Praxis the Journal of Ecopedagogy, 4*(1), 1–14.

Le Grange, L. (2016). Decolonising the University Curriculum: Leading Article. *South African Journal of Higher Education.* https://journals.co.za/content/high/30/2/EJC191701

Lotz-Sisitka, H. (2017). Decolonisation as Future Frame for Environmental and Sustainability Education: Embracing the Commons With Absence and Emergence. In P. Corcoran & J. Weakland (Eds.), *Envisioning Futures for Environment and Sustainability* (pp. 45–62). Wageningen Academic Publishers.

Lotz-Sisitka, H., Managi, S., Macintyre, T., Vickers, E. A., Chakraborty, A., Kitamura, Y., Xie, J., & Zhang, C. (2022). Global Change and the Contextual Dynamics Shaping Education: A View from the Sustainability-Education Nexus. In E. A. Vickers, K. Pugh & L. Gupta (Eds.), *Education and the Learning Experience, in Reimagining Education: The International Science and Evidence-based Education Assessment.* UNESCO MGIEP.

Orlovic Lovren, V., & Popovic, K. (2018). Lifelong Learning for Sustainable Development – Is Adult Education Left Behind? In W. L. Filho (Ed.), *Handbook of Lifelong Learning for Sustainable Development.* Springer International Publishing.

Macintyre, T., Chaves, M., Monroy, T., Zethelius, M. O., Villarreal, T., Tassone, V. C., & Wals, A. E. J. (2020). Transgressing Boundaries between Community Learning and Higher Education: Levers and Barriers. *Sustainability: Science Practice and Policy, 12*(7), 2601.

Maclean, R., Wilson, D., & Chinien, C. A. (2009). In R. Maclean & D. Wilson (Eds.), *International Handbook of Education for the Changing World of Work: Bridging Academic and Vocational Learning* (2009th ed.) [PDF]. Springer.

Mignolo, W. D. (2013). *The Darker Side of Western Modernity: Global Futures, Decolonial Options*. Duke University Press.

Mulà, I., & Tilbury, D. (2009). A United Nations Decade of Education for Sustainable Development (2005–14): What Difference Will It Make? *Journal of Education for Sustainable Development, 3*(1), 87–97.

Nguyen, P., Elliott, J. G., Terlouw, C., & Pilot, A. (2009). Neocolonialism in Education: Cooperative Learning in an Asian Context. *Comparative Education Review, 45*(1), 109–130.

Obondo, M. A. (2007). Tensions Between English and Mother Tongue Teaching in Post-Colonial Africa. In J. Cummins & C. Davison (Eds.), *International Handbook of English Language Teaching* (pp. 37–50). Springer US.

Pacific Forum Education Ministers (2006). *Pacific Education for Sustainable Development Framework*. United Nations.

Paden, M. (2002). *Drawing a Crowd*. World Resources Institute.

PCE (2004). *See Change: Learning and Education for Sustainability*. Parliamentary Commissioner for the Environment.

Pérez, J. G., & Llorente, M. T. P. (2005). Stultifera Navis: Institutional Tensions, Conceptual Chaos, and Professional Uncertainty at the Beginning of the Decade of Education for Sustainable Development. *Policy Futures in Education, 3*(3), 296–308.

Podger, D. (2009). Contributions of the American Bahá'í Community to Education for Sustainability. *Journal of Education for Sustainable Development, 3*(1), 65–74.

Ricard, M. (2013). *Trends and Issues of ESD in Europe* (特集 ESDの国際的な潮流). Trends and Issues of ESD in Europe. https://nier.repo.nii.ac.jp/record/2/files/kiyou142-101.pdf

Ryan, A., Tilbury, D., Corcoran, P. B., Abe, O., & Nomura, K. (2010). Sustainability in Higher Education in the Asia-Pacific: Developments, Challenges, and Prospects. *International Journal of Sustainability in Higher Education, 11*(2), 106–119.

Sauvé, L., & Berryman, T. (2005). Challenging a "Closing Circle": Alternative Research Agendas for the ESD Decade. *Applied Environmental Education and Communication, 4*(3), 229–232.

Selby, D. (2008). The Firm and Shaky Ground of Education for Sustainable Development. In D. S. J. Gray-Donald (Ed.), *Green Frontiers* (pp. 59–75). Sense Publishers.

Silova, I., Rappleye, J., & Auld, E. (2020). Beyond the Western Horizon: Rethinking Education, Values, and Policy Transfer. In G. Fan & T. S. Popkewitz (Eds.), *Handbook of Education Policy Studies: Values, Governance, Globalization, and Methodology, Volume 1* (pp. 3–29). Springer Singapore.

Spaargaren, G. (2011). Theories of Practices: Agency, Technology, and Culture: Exploring the Relevance of Practice Theories for the Governance of Sustainable Consumption Practices in the New World-Order. *Global Environmental Change: Human and Policy Dimensions, 21*(3), 813–822.

Sterling, S. (2004). Higher Education, Sustainability, and the Role of Systemic Learning. In P. B. Corcoran & A. E. J. Wals (Eds.), *Higher Education and the Challenge of Sustainability: Problematics, Promise, and Practice* (pp. 49–70). Springer Netherlands.

Sterling, S. (2005). *Linking Thinking, Education and Learning: An Introduction*. WWF-UK.

Sterling, S., & Huckle, J. (2014). *Education for Sustainability*. Routledge.

Stevenson, R. B. (2007). Schooling and Environmental Education: Contradictions in Purpose and Practice. *Environmental Education Research, 13*(2), 139–153.

Tavernaro-Haidarian, L. (2019). Decolonization and Development: Reimagining Key Concepts in Education. *Research in Education, 103*(1), 19–33.

Tilbury, D. (2003). The World Summit, Sustainable Development and Environmental Education. *Australian Journal of Environmental Education, 19*, 109–113.

Tilbury, D. (2004). Environmental Education for Sustainability: A Force for Change in Higher Education. In P. B. Corcoran & A. E. J. Wals (Eds.), *Higher Education and the Challenge of Sustainability: Problematics, Promise, and Practice* (pp. 97–112). Springer Netherlands.

Tilbury, D. (2012). Higher Education for Sustainability: A Global Overview of Commitment and Progress. In GUNI (Ed.), *Higher Education in the World* (pp. 18–22). Palgrave Macmillan.

Tilbury, D., & Cooke, K. (2005). *A National Review of Environmental Education and its Contribution to Sustainability in Australia: Frameworks for Sustainability – Key Findings.* Australian Government Department of the Environment and Heritage and Australian Research Institute in Education for Sustainability (ARIES).

Tilbury, D., Keogh, A., Leighton, A., & Kent, J. C. (2005). *A National Review of Environmental Education and Its Contribution to Sustainability in Australia: Further and Higher Education.* Australian Research Institute in Education for Sustainability (ARIES).

Tilbury, D., & Wortman, D. (2004). *Engaging People in Sustainability.* IUCN.

Tilbury, D., & Wortman, D. (2005). Whole School Approaches to Sustainability. *Geographical Education, 18*, 22–30.

Torres, R. M. (2004). *Lifelong Learning in the South: Critical Issues and Opportunities for Adult Education.* The Sida Studies-Series.

UNECE (2005). *UNECE Strategy for Education for Sustainable Development.* UN.

UNESCO (1975). *Belgrade Charter on Environmental Education.* UNESCO.

UNESCO (2002a). *Education for Sustainability. From Rio to Johannesburg: Lessons Learnt from a Decade of Commitment.* UNESCO.

UNESCO (2002b). *Johannesburg Declaration on Sustainable Development.* UNESCO.

UNESCO (2007). *The UN Decade of Education for Sustainable Development (DESD 2005-2014): The First Two Years.* UNESCO.

UNESCO (2021a). *Berlin Declaration on Education for Sustainable Development.* UNESCO. https://en.unesco.org/sites/default/files/esdfor2030-berlin-declaration-en.pdf

UNESCO (2021b). *Key Publications – ESD.* UNESCO. https://en.unesco.org/themes/education-sustainable-development/clearinghouse/publications

UNESCO-UNEP (1978). The Tbilisi Declaration. *Intergovernmental Conference on Environmental Education,* Tbilisi, October 14–26, 1977, Final Report. UNESCO/UNEP. http://unesdoc.unesco.org/images/0003/000327/032763eo.pdf.

United Nations. (2000). *Ten-year Review of Progress Achieved in the Implementation of the Outcome of the United Nations Conference on Environment and Development* (No. 55/199). United Nations.

Von Frantzius, I. (2004). World Summit on Sustainable Development Johannesburg 2002: A Critical Analysis and Assessment of the Outcomes. *Environmental Politics, 13*(2), 467–473.

Wals, A. E. J. (2009). A Mid-DESD Review: Key Findings and Ways Forward. *Journal of Education for Sustainable Development, 3*(2), 195–204.

Wals, A. E. J. (2012). *Shaping the Education of Tomorrow: 2012 Full-Length Report on The UN Decade Of Education For Sustainable Development.* UNESCO.

Wals, A. E. J. (2007a). Learning in a Changing World and Changing in a Learning World: Reflexively Fumbling Towards Sustainability. *Southern African Journal of Environmental Education, 24*(1), 35–45.

Wals, A. E. J. (2007b). *Social Learning Towards a Sustainable World: Principles, Perspectives, and Praxis.* Wageningen Academic Pub.

Walsh, C. (2010). Development as Buen Vivir: Institutional Arrangements and (de)colonial Entanglements. *Development, 53*(1), 15–21.

Wals, A. E. J., & Kieft, G. (2010). *Education for Sustainable Development: Research Overview,* 13. Sida. https://library.wur.nl/WebQuery/wurpubs/reports/402718

Wals, A., Pinar, W., Macintyre, T., Chakraborty, A., Johnson-Mardones, D., Waghid, Y., Tusiime, M., Le Grange, L., LL, Razak, D. A., Accioly, I., Xu, Y., Humphrey, N., Iyengar, R., Chaves, M., Herring, E., Vickers, E. A., Santamaria, R. D. P., Korostelina, K. V., & Pherali, T. (2022). Curriculum and Pedagogy in a Changing World. In E. A. Vickers, K. Pugh & L. Gupta (Eds.), *Education and Context in Reimagining Education: The International Science and Evidence Based Education Assessment.* UNESCO MGIEP.

WWF (2002). *Success Stories: Education and Capacity Building in Africa.* IUCN International.

6 Education and the Environment in the 2010s

Reframe and Transform Futures

Introduction: From Ad-Hoc to Systemic Educational Responses and Deep Learning

While each of the previous decades had their particular issues to deal with – from pollution and the threat of nuclear war, to globalisation and digitalisation – the 2010s saw a convergence of global concerns such as food, water, and energy shortages, bringing home the hard truth that time was running out to change humanity's path towards a more sustainable direction. In addition, climate change, which had been flagged already in 1992, in the United Nations Convention on Climate Change, and the Kyoto Protocol in 1997, became a major focal point for sustainability concerns at a governance level. This was due in part to the signing of the Paris Agreement in 2015 as well as heightened public concerns around the climate crisis (Mochizuki & Bryan, 2015; Reid, 2019; Stevenson et al., 2017).

The Stern review led to the consolidation of scientific predictions around the *economic* impacts of climate change and changed national policy discourses, triggering new thinking in relation to how environmental measures were to be financed (Stern, 2006). In addition, the focus on lifestyle changes and the need to develop more responsible forms of consumption remained strong during this decade. The 'Here and Now! Education for Sustainable Consumption' report, published by UNEP, provided policymakers with an instrument to understand the importance of education for sustainable consumption in supporting other policy goals, such as citizenship and democratic participation, environmental protection, and energy and climate policies (UNEP, 2010).

Another key document from this era was the UNESCO's Global Education Monitor (GEM) report, which concluded that education needed a major transformation to fulfil its potential and meet the current challenges facing humanity and the planet (UNESCO, 2016). Consolidating the original arguments made previously by critics such as Orr (1992) and Sterling (2001), the report pointed to dysfunctional economic systems which, it argued, are being mirrored by current education systems and practice. This

DOI: 10.4324/9781003467007-6

was an acknowledgement, for the first time at the intergovernmental level, that dominant paradigms in education were contributing to unsustainable development. It was no longer a question of 'adding on' sustainability to educational school curricula; instead, deep transformational change was needed (see Figure 6.1). Notably, in line with earlier UNESCO reports on ESD, it was the first GEM report to explicitly promote whole-institution and whole-school approaches.

The decade, thus, gave focus to educational arguments in support of whole-school approaches to sustainability that emerged in the 2000s and 2010s (see Henderson & Tilbury, 2004). The whole-school approach offered a way to create a connection between the taught and lived experience of the learner in schools (Tilbury, 2022), impacting on leadership and management, school buildings and grounds, formal and informal learning as well as the school's relationship with the local community (Mathie & Wals, 2022). Ultimately, it sought to connect and make spaces for students' interests from outside of the classroom, with what they were learning at school. In his text, 'Sustainable Schools; Sustainable Futures' David Hicks (2012) argues that whole-school approaches are not just about joining up local practices; they should also make connections with global communities and draw attention to 'futures' perspectives that generate creative and critical responses and help young people tackle complex sustainability challenges.

The DESD Comes to an End

The 2010s also saw the end of the DESD in 2014, which though contentious in its design and engagement, had a big impact on the framing of education and learning for the environment and sustainability. For example, the DESD carved out a prominent role for higher education in the global vision and pathway for sustainable development. It achieved this through successfully raising the profile of its efforts across campuses and encouraging the embedding of ESD into core curricula and developing graduates that were literate in sustainability (Sterling et al., 2013). This all happened while igniting international debate about how higher education could best, and more directly, accelerate change across local to global communities (Nomura & Abe, 2010; Tilbury, 2013), and preparing the sector for the arrival of the SDGs and SDG impact assessments later in the decade.

It is important to note, however, that many of the innovations that happened during this decade occurred on the margins of the sector. As argued by Tilbury, the 'rebooting' of higher education towards sustainable development had not yet taken place (Tilbury, 2014). This critical sentiment is shared by Huckle and Wals (2015), who consider the DESD to have failed to dent the 'business as usual' approach in schools, colleges, and universities. The authors attributed this to education's inability to acknowledge or challenge

The GEM report *Education for people and planet: Creating sustainable futures for all* (UNESCO, 2016) signalled a change of tone and narrative in terms of the role of education in bringing about more sustainable futures. In italics below, we highlight conclusions and recommendations from the report that signal a more critical voice in UN circles with regards to conventional economic thinking, but also the ones that speak to overcoming inequality and injustice, and the recognition of indigenous voices and ways of being.

- *Current models of economic growth cause environmental destruction*
- For education to be transformative in support of the new sustainable development agenda, 'education as usual' will not suffice.
- *Education cannot fight inequality on its own. Labour markets and governments must not excessively penalise lower income individuals. Cross-sectoral cooperation can reduce barriers to gender equality.*
- A whole-school approach is needed to build green skills and awareness. Campaigns, companies, as well as community and religious leaders must advocate for sustainability practices. Non-formal education and research and development should also help solve global environmental challenges.
- *Expand education on global citizenship, peace, inclusion, and resilience to conflict. Emphasise participatory teaching and learning especially in civic education. Invest in qualified teachers for refugees and displaced people, and teach children in their mother language. Incorporate education into the peacebuilding agenda.*
- Distribute public resources equitably in urban areas, involving the community in education planning.
- *Mobilise domestic resources, stop corporate tax evasion, and eliminate fossil fuel subsidies to generate government revenue for fundamental needs such as education and health.*
- Include education in all discussions on urban development. Improve and fund urban planning programmes and curricula to include cross-sector engagement and develop locally relevant solutions.
- *Promote the value of indigenous livelihoods, traditional knowledge and community-managed or -owned land through actions such as land conservation and locally relevant research.*
- Engage community elders in curricular development and school governance, produce appropriate learning materials, and prepare teachers to teach in mother languages.
- *Incentivize universities to support the development of graduates and researchers who address large-scale systemic challenges through creative thinking and problem-solving.*
- Promote cooperation across all sectors to reduce policy-related obstacles to full economic participation by women or minority groups, as well as discrimination and prejudice that also act as barriers.
- *Support multistakeholder governance for the sustainable management of natural resources and of public and semi-public rural, urban, and peri-urban spaces.*

Figure 6.1 The Global Education Monitoring (GEM) Report 2016

neoliberalism – a hegemonic force blocking transitions towards genuine sustainability. Huckle and Wals (2015) instead suggest that global education for sustainability citizenship (GESC) would have provided a more realistic focus for such an initiative, as there was too little focus on power, politics, and citizenship in the DESD. Furthermore, while acknowledging the importance of multiple perspectives and dialogue within ESD (e.g. Jickling & Wals, 2008), Kopnina (2014) warns that the anthropocentric agenda of ESD may in fact be counterproductive to the efficacy of environmental learning and education in fostering a citizenry that is, as stated by the Belgrade Charter, 'aware of, and concerned about, the environment and its associated problems' (UNESCO, 1975). Kopnina notes that if learners do not become aware of the deep interlinkages between humanity and nature, then what they are learning may simply continue reproducing the existing status quo, instead of leading to necessary transformations in society. What is important is for learners to understand that daily choices related to how we choose to travel, for example, have implications for people and planet but also that individual choices are heavily influenced by the systems and structures in which they are immersed (Tilbury, 2011).

The Arrival of the Sustainable Development Goals (SDGs)

Following on from the DESD, in January 2016, the 17 Sustainable Development Goals (SDGs) of the 2030 Agenda for Sustainable Development officially came into force. The adoption by world leaders of the SDGs in the 2015 historic UN Summit, though not legally binding, requires governments to take ownership and mobilise efforts to fight poverty, inequality and tackle climate change, while ensuring that no one is left behind. Member states have the responsibility to establish national frameworks for the achievement of the 17 Goals and review implementation progress. ESD is explicitly stated in SDG 4 on quality education, in target 4.7:

> By 2030, ensure that all learners acquire the knowledge and skills needed to promote sustainable development, including, among others, through Education for Sustainable Development and sustainable lifestyles, human rights, gender equality, promotion of a culture of peace and non-violence, global citizenship and appreciation of cultural diversity and of culture's contribution to sustainable development.

UNESCO's Global Action Programme (GAP) on ESD, the follow-up programme to the DESD, was launched in November 2014 in Aichi-Nagoya and sought to scale-up education efforts to accelerate progress towards sustainable development. The GAP focused on five priority action areas:

1) to advance policy; 2) transform learning and training environments; 3) build the capacities of educators and trainers; 4) empower and mobilise youth, and 5) accelerate sustainable solutions at a local level (UNESCO, 2020). Stratford and Wals (2020) point out that in order to be successful in all these areas, policy environments will be needed that are conducive to such actions.

Higher Education and the SDGs

The higher education sector was quick to embrace the SDG responsibilities with university leaders, student bodies and sector networks committing to practical steps to advance the SDGs (Chankseliani & McCowan, 2021; Tilbury, 2014). This readiness to engage came as no surprise, given that many higher education institutions were already committed to sustainable development issues in the 1980s and 1990s and had experimented with some of the opportunities offered by DESD (Abe, 2009; Corcoran & Wals, 2004; Jones et al., 2010; Tilbury, 2014). In the words of Ryan and Tilbury (2013b), the higher education sector had 'dipped its toes' into the sustainable development waters, with awards and case studies of good practice documenting the diversity of efforts and small steps taken forward in this agenda. At the same time, the authors noted how research journals, evaluation reports, and rankings pointed to how early pioneers were meeting substantial obstacles in mainstreaming sustainability pilot projects, sustaining the impacts of their efforts, and embedding change into higher education systems.

Although a necessary step in the process of social and institutional change, the involvement of disciplinary experts and broader engagement of higher education practitioners with limited background on pedagogical processes or learning for sustainability has resulted in obstacles to implementing sustainability into educational contexts (Tilbury, 2013). The issue was, and continues to be, that the SDGs are being primarily interpreted as thematic (i.e. as topics to be added to the existing offerings) and not as doorways to revisit or review existing offerings. These practices therefore fail to challenge educational practice in higher education and thus continue to reproduce unsustainable visions of the future (Barth et al., 2015; Tilbury, 2012). At the same time, much has been learnt over the years about the challenges and obstacles that accompany sustainable development initiatives as well as the resilience needed to progress the ambition to another level (Barlett & Chase, 2013; Tilbury, 2019).

Converging Educational Streams

Another educational response is the recognition that global and local issues are inextricably intertwined and addressing them requires collective action (Hicks, 2012). This sees a convergence of different planetary adjectival education that have been around since the 1980s, such as health education, peace education, human rights education, and biodiversity education, in addition to newer ESD frames. Each of these educations addresses some planetary issue, related to citizenship, health, wellbeing, good care of the environment, and protection of species. What we see in the 2010s is that these issues are very much interconnected and deeply entangled. They require what we might call boundary crossing (Akkerman & Bakker, 2011; Fortuin, 2015), as well as the interdisciplinarity of past decades, as well as transdisciplinarity (Keitsch & Vermeulen, 2020). We can see this in the blurring between formal, informal, and non-formal learning, between genders, between generations, between disciplines, between sectors of society, and between different knowledge systems (e.g. scientific, indigenous, and local/experiential (Cutter-Mackenzie-Knowles et al., 2020)). Adjectival education areas also shared a common interest in 'flexible pedagogies' (Ryan & Tilbury, 2013a) that have the potential to challenge power relationships in education and build the capabilities of learners so that they can more actively engage in change for sustainability.

Another interesting convergence is between science education and Environmental Education. As discussed by Wals and colleagues (2014), science education, which focuses on understanding natural systems and processes through teaching knowledge and skills, has traditionally been disconnected to Environmental Education, which explores the sociopolitical, value-laden, place-based, and emotional contexts in which environmental issues take place. The authors drew attention to citizen science – which had shown to be an effective approach to engaging people, scientists, and local communities with science on relevant environmental issues in place-based contexts (Bonney et al., 2014; Dillon et al., 2016). Once again, in this decade, we see educational movements started in the 1980s and 1990s, taking root in mainstream practice related to environmental learning and education.

Early Childhood Education and Care

In the early 2010s, UNESCO published its first international work linking Early Childhood Education and Care (ECEC) together with sustainability (UNESCO, 2012). During this time, research in early childhood was beginning to argue for ESD not only as content but also as a way of teaching to children, though this is controversial, with an opposing view holding that children should be sheltered from global problems (Elliott & Davis, 2009; Pearson & Degotardi, 2009; Pramling Samuelsson, 2011). Noteworthy is the 68th World Organisation for Early Childhood Education (OMEP) World Assembly held in Seoul, Korea, which focussed on sustainability in ECEC. Several speakers and papers emphasised the innate power of children to see the world as it unfolds in a more holistic and relational way but once they enter the world of education, they lose this power and ability. As noted in the introductory paper in the special issue of the *International Journal of Early Childhood Education*, connected to the meeting: 'What seems critical is that children encounter a multiplicity of different worlds by crossing boundaries, both individually and together, and having bodily experiences that strengthen their *relationality* with the human, the non-human and the material. It is through these encounters that agency, care and empathy can develop. All three of these qualities are foundational for a world that is more sustainable than the one currently in prospect' (Wals, 2017, p. 162).

STEM, ESTEM, and STEAM

In North America, the STEAM (Science, Technology, Engineering, the Arts, and Mathematics) movement in science education helped learning move away from segmented subject domains in science teaching. This was an attempt to make learning more appealing and relevant to students, whereby these efforts coincided with a call for developing 21st-century skills like collaboration, questioning, problem-solving, and critical thinking (Kivunja, 2014).

Referring back to the earlier mentioned Sputnik era, President Barack Obama, in his 2011 State of the Union, described this decade as 'our generation's Sputnik moment'. It was a plea for the USA to invest in science education, not so much to respond to environmental and sustainability issues, but rather to prevent the country from falling behind in technological innovation and to stay competitive with other nations (Gunn, 2017). Later, another 'E' for Environment was added to recognise the importance of not just focusing on innovation in education but also on environmental conservation and emerging sustainability challenges (Gupta et al., 2018). As a further acknowledgement that inter- and transdisciplinarity and boundary crossing between subject areas needed more attention and space in education some groups advocated for adding the 'A' of the arts (and humanities) to make STEAM. Smith and Watson (2019) have done a helpful analysis of framings of STEM and STEM education in relation to Education for Sustainability (EfS). While the STEM movement did not have a major effect on the field of Environmental Education and ESD, there have been similar tendencies in science education mainly to enrich the content with environmental and sustainability-related topics like plastic soup, climate change, alternative energy sources, and to make connections with other subjects where possible. Examples of green or sustainable chemistry, physics, and mathematics education are plentiful (e.g. Jegstad and Sinnes (2015) for Chemistry Education, Nowotny et al. (2018) for Physics Education in relation to energy, and Renert (2011) for Mathematics Education).

In countries where 'subject matter didactics' play a key role in education, like in many Northern European countries, there were initiatives to develop and introduce approaches to learning that could help students in dealing with the complex and sometimes controversial and ambiguous nature of sustainability issues as they often emerge in geography, language and science classrooms. Often these approaches focussed on dialogue, changing perspectives, reflexivity, and making connections (Öhman & Sund, 2021; Sjöström & Rydberg, 2018; Van Poeck, 2019), which align with methodologies that the DESD had been advocating for the last decade (Tilbury, 2011). It was during this time that the prestigious magazine *Science* published an article noting an unhealthy relationship between environmental and science education (Wals et al., 2014). The authors characterise the relationship as 'distant, competitive, predator prey and host-parasite' and call for a convergence between environmental and science education to assure young people become meaningfully engaged in socio-ecological issues, whereby citizen science is put forward as key to bridging the divide.

The Rise of Relational Approaches to Teaching and Learning

What became increasingly clear in the 2010s was that to remain relevant, education needed to be reconfigured to contribute to the sustainability movement in society and to restore and regenerate relations and connections between people, places, and the more-than-human world. This is the task of all education, from early childhood education and care to vocational, higher, and continuing education. Systems thinking (Dunnion & O'Donovan, 2014) but also notions of entanglement (Hofverberg, 2020; Verlie, 2017; Wessels et al., 2022) begin to take a more prominent role in education, requiring that we see connections and interdependencies, and learn to see ourselves as part of a system.

Reminding us of the need to address anthropocentric concerns over dominant development paradigms, there are those who argue that we should connect with different species as well, moving towards post-human perspectives (e.g. Malone et al., 2020). In this way, we decentre the human and become more ecocentric, biocentric, and less anthropocentric, so as to address the complex nature of current sustainability challenges, and the need for citizens who can adequately respond to them (Lloro-Bidart, 2018). These perspectives have been around for a long time historically, and had early roots in systems thinking approaches (see Stephen Sterling's work on 'linking thinking' in chapter 5), but received more societal traction in the 2010s. In addition, the rise of the earlier mentioned arts-based 'immersive' approaches also represents this relationality but more from an embodied and emotional perspective compared to the more rational systems thinking perspective. For a review, interpretation, and enactment of arts-based approaches to education, the work of van Boeckel provides a good starting point (van Boeckel, 2014).

Translated to education we see resource materials for teachers on alternative economic models becoming available, such as on the website *The Doughnut Economics Action Lab* (2024). This website provides open access tools that educators can use to turn Doughnut Economics 'from a radical idea into transformative action'. In addition to case studies, the website supports an online community of educators, policymakers, and community members that can share resources and information so as to turn Doughnut Economics ideas into practice.

Figure 6.2 Teaching doughnut economics.

Source: Raworth (2017)

A different relational angle comes from emerging ideas within economic thinking that emphasise circularity, closing cycles, and considering the whole life cycle and production chain of products. Underlying this thinking lies an attempt to reconfigure the system towards regenerative and distributive economies. These approaches aim to balance the needs of humanity within planetary limits, thereby becoming less dependent on global capitalist and neo-colonial extractive models (Morseletto, 2020). Some examples are doughnut economics (Raworth, 2017) (see Figure 6.2), the circular economy (Kirchherr et al., 2017), as well as degrowth (Schneider et al., 2010).

Decolonising Ways of Thinking

Another response to the effects of the neoliberal ideology on the educational system (Apple, 2013) is the interest in decolonisation education. This strand examines the limitations and biases of curriculum and teacher education and training, and the social, political, and environmental legacies of colonisation, and how they have influenced education policies. Like other critical strands, the decolonial current highlights how the sustainability-through-growth paradigm increases inequalities and links biodiversity loss, climate change, and social tensions. Kothari and colleagues (2020), for example, note that SDG 4 (quality education) adopts a particularly dominant view of education and development, which needs to be deeply critiqued in terms of the skills which are to be learnt and taught. This requires what Sund and Pashby (2020) describe as 'delinking as a decolonial praxis,' for example, through the exploration of multiple perspectives that reflect different worldviews and narratives and explore and engage with the complexities and contractions between them.

While decolonial education has been around for decades, what we begin to see in the 2010s is that the decolonial approach finds more clarity and relevance in educational circles through movements such as the 2015 #DecoloniseTheCurriculum movement, and the 2016 #FeesMustFall in South Africa. These movements brought students to the streets and resulted in public discussions on curriculum renewal and decoloniality. It also drew attention to land education issues and their links to place-based Environmental Education (Tuck et al., 2014). While Le Grange and colleagues (2020) note the importance of the decolonisation question and provide alternatives to the Western imposed outcomes-based approach used in South African higher education, the authors also note concerns of institutions turning to instrumentalist and quick-fix solutions to decolonise curricula, which can result in decolonial-washing rather than transformative change.

There is also increasing interest in areas such as Indigenous Knowledge Systems (IKS) and indigenous Environmental Education, which conceptualise knowledge as holistic, organic, and relational, being made up of connections to living and non-living beings and entities (Kayira et al., 2022). From a teaching and learning for the environment perspective, such decolonised models of education are rooted in connections to place and empower students to establish links to their local community, striving to restore local, traditional, and cultural knowledge. Rather than forcing one dominant ideology on students, decolonisation education promotes intercultural and inclusive learning, recognising pluriversal ways of being in the world. Some authors highlight the critical analysis and reflexivity involved in engaging with alternative development models that question taken-for-granted assumptions and suggest alternatives (Kopnina, 2020; Kothari et al., 2020; see Figure 6.3). While such alternative paradigms were taken more seriously in the 2010s, others call for increased attention in education to include indigenous and black feminist approaches, alongside renewed attention to social justice and indigenous systems of knowledge, with a territorial understanding and focus on education (Maina-Okori et al. (2018).

Buen Vivir (integrative and collective well-being) is based on indigenous forms of knowledge, critical intellectuals, and political movements in Latin America (Cortina & Earl, 2021; Gudynas, 2011). It is a multidimensional and plural concept, which presents a fundamental challenge to the modern development paradigm, including an educational system which is complicit with current economic models (Brown & McCowan, 2018). Buen vivir has also made its way into international documents recognising other knowledge systems (UNESCO, 2016). Another example is the traditional concept of Sejahtera, which is a philosophy of sustainable living and balanced coexistence in the Malay language (Razak, 2018), while Ubuntu is an African concept encapsulated in the proverb 'I am because you are,' and ecological Swaraj, encapsulates radical ecological democracy in India (Kothari et al., 2014). While these alternative models have their own cultural and historical contexts, they share common characteristics, such as relationality, plurality, and respect and care for the earth.

Figure 6.3 Alternative development models, relationality, and plurality

Maturation and Expansion of Research on Environmental and Sustainability Education and Practice

Over the years, the field of Environmental Education research has matured and expanded. First, there has been enormous success of special interest groups (SIG) on environmental and sustainability education. We can see this in major international educational research networks like the American Educational Research Association (AERA) which in the early 1990s established a SIG on Environmental and Ecological Education, and the European Conference on Educational Research (ECER). Second, the AERA, by far the world's biggest educational research network, begins to recognise the importance of this research niche within educational research. In part fueled by the SIG on Ecological and Environmental Education, we can see this in AERA sponsoring the publication of a handbook focusing entirely on Environmental Education research, but also from the publication of the *International Handbook on Environmental Education Research* (Stephenson et al., 2013). Over the years the research field has become far more international and diverse compared to the early years which were dominated by white, male, and Western thinkers with a mostly positivist and scientific frame.

This diversity also expressed itself in a broadening of research perspectives and methodologies (Nomura, 2017). During this decade, Dillon and Wals (2006) made a distinction between research as mining, research as learning, and research as activism, with the latter gaining prominence in more recent decades. Lotz-Sisitka and colleagues of the international T-Learning Project on Transformations to Sustainability introduced the idea of co-designing transgressive research that consciously attempts to break with normalised problematic patterns (Lotz-Sisitka et al., 2016). Work undertaken to embed sustainability learning in higher education has also given rise to joint efforts, and a critical alignment, between organisational development, professional development, and research approaches for sustainability (Tilbury et al., 2004). In the same vein, *The University Educators for Sustainable Development* (*UE4SD*) initiative brought together 45 universities across Europe to explore how research could be a gateway for institutional change for sustainability in higher education (Tilbury et al., 2012)

Mackenzie and colleagues from the Sustainability Education Policy Network (SEPN) and the subsequent Monitoring and Evaluation of Climate Communication and Education (MECCE) project, have conducted policy-related research that started during this decade and has continued ever since. These two major projects can be seen as examples of policy-related research that has not only influenced UNESCO circles but also policymaking at the national level in some countries (see Rickinson and McKenzie, 2021). In addition, we now also witness the introduction of arts-based methods in environmental and sustainability education research, especially in the context of establishing human-nature connections (Eernstman & Wals, 2013; Eernstman et al., 2012; Muhr, 2020).

Quality Education Frames and Perspectives

The SDGs introduced the language of 'Quality education' in conceptions of learning for sustainability, continuing to challenge educators to consider learner processes and outcomes and not just content or themes associated with this agenda. This development drew the attention of quality education professionals (e.g. inspectors, reviewers, curriculum experts, and peer assessors) whose task is to assure the quality of learning processes in schools, colleges, and universities. These professionals serve as powerful influencers in education and brought new vocabulary and energy as well as frameworks to this decade that helped mainstream environmental and sustainability learning, particularly in higher education.

In 2013, funded by the Quality Assurance Agency (QAA) for Higher Education in England, Tilbury and Ryan developed a resource for leading institutional change for sustainable development from a quality education perspective (Tilbury et al., 2012). The guide combined lessons from the five institutional pilot projects with a sector-wide view of how ESD connects with quality assurance and enhancement in higher education. A year later, and building on from this work, the QAA convened an advisory group to develop national guidelines on ESD with the ultimate aim of supporting its inclusion into the *UK Quality Code for Higher Education*. This was a significant development as the Code informs national audits of curriculum standards and enhancement systems for all UK universities (QAA, 2021). Shortly after these, the International Quality Assurance Agency for Higher Education (INQAAHE) funded an initiative that sought to embed SDGs into national and regional quality systems and resulted in an indicator framework that was formally adopted by the participating agencies in Andorra and Spain as a means to assess quality in ESD offerings in higher education (Tilbury et al., 2019). The framework proposed a whole-institution response to embedding ESD into higher education institutions and marked tangible progress markers. This work, in turn, influenced the UNECE's strategic framework for ESD which adopted 'quality education' as a key strand of commitment for the next decade (UNECE, 2023).

On the other side of the globe, an initiative led by the University of Newcastle Australia, and Association of Deans of Business Schools in Australia, wrestled with similar questions but generated an alternative way forward. The 'National Learning and Teaching Standards for Environment and Sustainability' project (Phelan et al., 2015) was funded by the Australian Learning and Teaching Commission and defined what students need to know and be able to do upon graduation, often referred to as student sustainability graduate attributes or competencies. Led by Bonnie McBain and Liam Phelan, the project team consulted with a wide range of stakeholders including tertiary educators, quality experts and researchers, employers, indigenous communities, and practitioners and students (Phelan et al., 2015). Significantly, the project resulted in the inclusion of sustainability in base disciplinary knowledge and subject benchmarks, helping to ensure institutions offer students opportunities to gain practical experience and skills, as well as the embedding of sustainability into the Australian Qualification Framework (McBain et al., 2024). These quality education initiatives served as a powerful mechanism for mainstreaming environmental and sustainability learning in higher education, but mostly in the countries where the groundwork had been undertaken.

Moving from the 2010s into the 2020s

The 2010s saw a convergence of educational streams around the drive towards addressing sustainability concerns, which only seemed to be increasing in societies around the world. The SDGs strengthened sustainability narratives and triggered education institutions and systems to consider the implications of environment, climate change, and social injustices in the curriculum. There was also much more of an effort to embed sustainability principles and practices into schools, as seen in the whole-school approach which became dominant in education. In this way, the decade can be seen as a reframing of education towards a transformative outlook, rather than the focus in the previous decades on adapting to external environmental pressures. More critical perspectives began to consolidate, also within the United Nations, calling for a rethinking of growth-oriented economic thinking. This saw a move towards more distributive and circular economic thinking that is more in tune with ecology and social justice. In line with this, some organisations and NGOs started focussing on education for sustainable lifestyles and alerting consumers to the 'power of the purse'. At the same time, the human–nature relationship, the recognition of local knowledge, and the regeneration of indigenous knowledge become more prominent. In light of these shifts, there is an increased recognition that the modern educational system needs to be transformed to create space for boundary crossing, decolonising voices, critical thinking, connecting with nature, and dialogue about alternative development models. Along with this, there is a call for new ways of teaching and learning, based on principles of relationality, plurality, and respect and care for the earth (See Figure 6.4).

The following 2020s saw these strands continue to question the role of education and learning in the quest for a sustainable planet. This is particularly apparent in concerns such as climate change – and associated climate action, climate justice, and climate anxiety – that underscore the urgency of integrating the perspectives of youth and marginalised communities into both environmental discourse and educational practices. As the following chapter will show, while there is a commendable global acknowledgment of the urgent need for environmental engagement and action, the effectiveness of educational initiatives to contribute to a more sustainable, inclusive, and healthy world remains an open question (Figure 6.4).

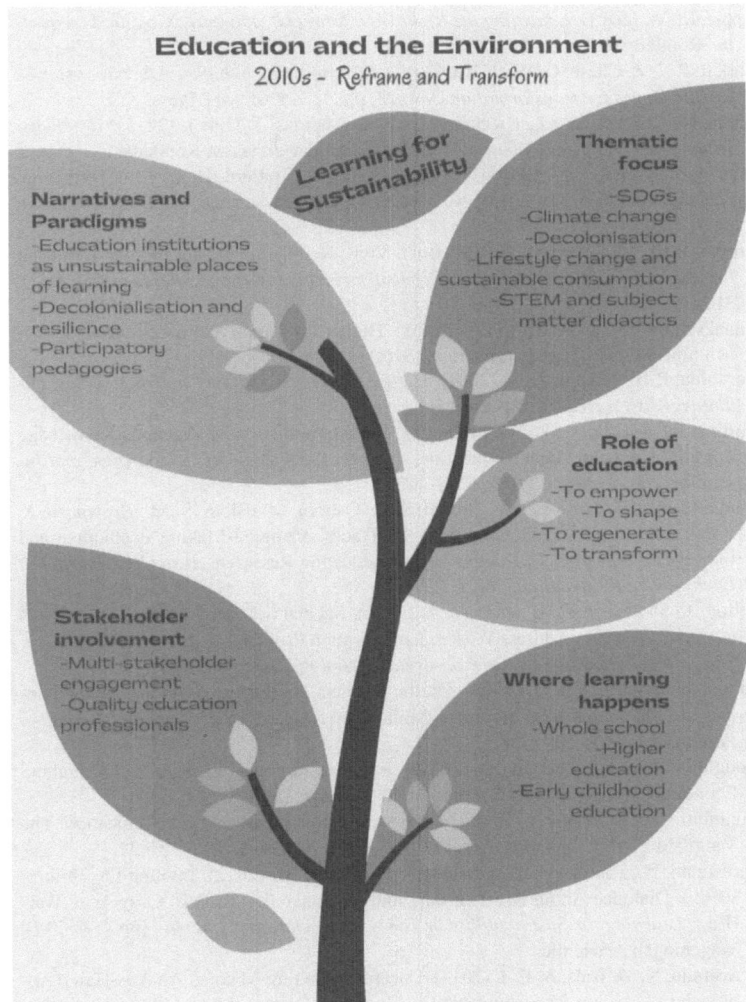

Figure 6.4 Summary of education and the environment in the 2010s

References

Abe, O. (2009). Current Status and Perspectives of Education for Sustainable Development (ESD). *Japanese Journal of Environmental Education, 19*(2), 21–30.

Akkerman, S. F., & Bakker, A. (2011). Boundary Crossing and Boundary Objects. *Review of Educational Research, 81*(2), 132–169). https://doi.org/10.3102/0034654311404435

Apple, M. W. (2013). *Educating the Right Way: Markets, Standards, God, and Inequality*. Routledge.

Barlett, P. F., & Chase, G. W. (Eds.). (2013). *Sustainability in higher education: stories and strategies for transformation* (Vol. 23, pp. 485–496). MIT Press.

Barth, M., Michelsen, G., Rieckmann, M., & Thomas, I. (Eds.). (2015). *Routledge Handbook of Higher Education for Sustainable Development*. Routledge.

Bonney, R., Shirk, J. L., Phillips, T. B., Wiggins, A., Ballard, H. L., Miller-Rushing, A. J., & Parrish, J. K. (2014). Next Steps for Citizen Science. *Science, 343*(6178), 1436–1437.

Brown, E., & McCowan, T. (2018). Buen Vivir: Reimagining Education and Shifting Paradigms. *Compare: A Journal of Comparative and International Education, 48*(2), 317–323.

Chankseliani, M., & McCowan, T. (2021). Higher Education and the Sustainable Development Goals. *Higher Education, 81*(1), 1–8.

Corcoran, P. B., & Wals, A. E. (2004). *Higher Education and the Challenge of Sustainability*. Kluwer Academic Press.

Cortina, R., & Earl, A. (2021). Embracing Interculturality and Indigenous Knowledge in Latin American Higher Education. *Compare: A Journal of Comparative and International Education, 51*(8), 1208–1225.

Cutter-Mackenzie-Knowles, A., Brown, S. L., Osborn, M., Blom, S. M., Brown, A., & Wijesinghe, T. (2020). Staying-with the Traces: Mapping-Making Posthuman and Indigenist Philosophy in Environmental Education Research. *Australian Journal of Environmental Education, 36*(2), 105–128.

Dillon, J., Stevenson, R. B., & Wals, A. (2016). Special Section: Moving from Citizen to Civic Science to Address Wicked Conservation Problems. *Conservation Biology: The Journal of the Society for Conservation Biology, 30*(3), 450–455.

Dillon, J., & Wals, A. E. J. (2006). On the Dangers of Blurring Methods, Methodologies and Ideologies in Environmental Education Research. *Environmental Education Research, 12*(3/4), 549–558.

Doughnut Economics Action Lab (2024). *Doughnut Economics Action Lab*. Doughnut Economics Action Lab. https://doughnuteconomics.org/

Dunnion, J., & O'Donovan, B. (2014). Systems Thinking and Higher Education: The Vanguard Method. *Systemic Practice and Action Research, 27*(1), 23–37.

Eernstman, N., van Boeckel, J., Sacks, S., & Myers, M. (2012). Inviting the Unforeseen: a Dialogue About Art, Learning and Sustainability. In P. B. C. A. E. J. Wals (Ed.), *Learning for Sustainability in Times of Accelerating Change* (pp. 201–212). Wageningen Academic.

Eernstman, N., & Wals, A. E. J. (2013). Locative Meaning-Making: An Arts-Based Approach to Learning for Sustainable Development. *Sustainability, 5*(4), 1645–1660.

Elliott, S., & Davis, J. (2009). Exploring the Resistance: An Australian Perspective on Educating for Sustainability in Early Childhood. *International Journal of Early Childhood, 41*(2), 65–77.

Fortuin, K. P. J. (2015). In R. Leemans (Ed.), *Heuristic Principles to Teach and Learn Boundary Crossing Skills in Environmental Science Education*. Wageningen University. https://library.wur.nl/WebQuery/wurpubs/fulltext/356213

Gudynas, E. (2011). Buen Vivir: Today's Tomorrow. *Development, 54*(4), 441–447.

Gunn, J. (2017, November 3). *The Evolution of STEM and STEAM in the U.S.* Resilient Educator. https://resilienteducator.com/classroom-resources/evolution-of-stem-and-steam-in-the-united-states/

Gupta, R., LaMarca, N., Rank, S. J., & Flinner, K. (2018). The Environment as a Pathway to Science Learning for K–12 Learners—A Case Study of the E-STEM Movement. *Ecopsychology, 10*(4), 228–242.

Henderson, K., & Tilbury, D. (2004). *Whole-School Approaches to Sustainability: An International Review of Sustainable School Programs.* Australian Research Institute in Education for Sustainability (ARIES): Australian Government.

Hicks, D. (2012). *Sustainable Schools, Sustainable Futures.* Godalming: WWF. https://www.risingstars-uk.com/media/Rising-Stars/Series%20Images/Voyagers%20Free%20Samples/Sustainable-Schools,-Sustainable-Futures-Contents.pdf

Hofverberg, H. (2020). Entangled Threads and Crafted Meanings – Students' Learning for Sustainability in Remake Activities. *Environmental Education Research, 26*(9–10), 1281–1293.

Huckle, J., & Wals, A. E. J. (2015). The UN Decade of Education for Sustainable Development: Business as Usual in the End. *Environmental Education Research, 21*(3), 491–505.

Jegstad, K. M., & Sinnes, A. T. (2015). Chemistry Teaching for the Future: A Model for Secondary Chemistry Education for Sustainable Development. *International Journal of Science Education, 37*(4), 655–683.

Jickling, B., & Wals, A. E. J. (2008). Globalization and Environmental Education: Looking Beyond Sustainable Development. *Journal of Curriculum Studies, 40*(1), 1–21.

Jones, P., Selby, D., & Sterling, S. (Eds.). (2010). *Green Infusions: Embedding Sustainability across the Higher Education Curriculum.* Earthscan.

Kayira, J., Lobdell, S., Gagnon, N., Healy, J., Hertz, S., McHone, E., & Schuttenberg, E. (2022). Responsibilities to Decolonize Environmental Education: A Co-Learning Journey for Graduate Students and Instructors. *Societies, 12*(4), 96.

Keitsch, M. M., & Vermeulen, W. J. V. (2020). *Transdisciplinarity For Sustainability: Aligning Diverse Practices.* Routledge.

Kirchherr, J., Reike, D., & Hekkert, M. (2017). Conceptualizing the Circular Economy: An Analysis of 114 Definitions. *Resources, Conservation and Recycling, 127*, 221–232.

Kivunja, C. (2014). Do You Want Your Students to Be Job-Ready With 21st Century Skills? Change Pedagogies: A Pedagogical Paradigm Shift from Vygotskyian Social Constructivism to Critical Thinking, Problem Solving and Siemens' Digital Connectivism. *International Journal of Sustainability in Higher Education, 3*(3), 81–91.

Kopnina, H. (2014). Revisiting Education for Sustainable Development (ESD): Examining Anthropocentric Bias Through the Transition of Environmental Education to ESD. *Sustainable Development, 22*(2), 73–83.

Kopnina, H. (2020). Education for the Future? Critical Evaluation of Education for Sustainable Development Goals. *The Journal of Environmental Education, 51*(4), 280–291.

Kothari, A., Demaria, F., & Acosta, A. (2014). Buen Vivir, Degrowth and Ecological Swaraj: Alternatives to Sustainable Development and the Green Economy. *Development, 57*(3), 362–375.

Kothari, A., Salleh, A., Escobar, A., Demaria, F., & Acosta, A. (Eds.). (2020). *Pluriverse: A Post-Development Dictionary* (pp. 379–381). Tulika.

Le Grange, L., Du Preez, P., Ramrathan, L., & Blignaut, S. (2020). Decolonising the University Curriculum or Decolonial-Washing? A Multiple Case Study. *Journal of Education (University of KwaZulu-Natal), 80*, 25–48.

Lloro-Bidart, T. (2018). A Feminist Posthumanist Ecopedagogy in/for/with Animalscapes. *The Journal of Environmental Education, 49*(2), 152–163.

Lotz-Sisitka, H., Ali, M. B., Mphepo, G., Chaves, M., Macintyre, T., Pesanayi, T., Wals, A. E. J., Mukute, M., Kronlid, D., Tran, D. T., Joon, D., & McGarry, D. (2016). Co-Designing Research on Transgressive Learning in Times of Climate Change. *Current Opinion in Environmental Sustainability, 20*, 50–55.

Maina-Okori, N. M., Koushik, J. R., & Wilson, A. (2018). Reimagining Intersectionality in Environmental and Sustainability Education: A Critical Literature Review. *The Journal of Environmental Education, 49*(4), 286–296.

Malone, K., Tesar, M., & Arndt, S. (2020). *Theorising Posthuman Childhood Studies.* Springer Singapore.

Mathie, R. G., & Wals, A. E. J. (2022). *Whole School Approaches to Sustainability: Exemplary Practices from around the World.* Education & Learning Sciences/ Wageningen University.

McBain, B., Phelan, L., Ferguson, A., Brown, P., Brown, V., Hay, I., Horsfield, R., Taplin, R., & Tilbury, D. (2024). Collaboratively Crafting Learning Standards for Tertiary Education for Environment and Sustainability. *International Journal of Sustainability in Higher Education, 25*(2), 338–354.

Mochizuki, Y., & Bryan, A. (2015). Climate Change Education in the Context of Education for Sustainable Development: Rationale and Principles. *Journal of Education for Sustainable Development, 9*(1), 4–26.

Morseletto, P. (2020). Restorative and Regenerative: Exploring the Concepts in the Circular Economy. *Journal of Industrial Ecology, 24*(4), 763–773.

Muhr, M. M. (2020). Beyond Words – The Potential of Arts-Based Research on Human-Nature Connectedness. *Ecosystems and People, 16*(1), 249–257.

Nomura, K. (2017). Environmental Education Research in Japan—A Fragmented Field of Inquiry. *Japanese Journal of Environmental Education, 26*(4), 4_57–64.

Nomura, K., & Abe, O. (2010). Higher Education for Sustainable Development in Japan: Policy and Progress. *International Journal of Sustainability in Higher Education, 11*(2), 120–129.

Nowotny, J., Dodson, J., Fiechter, S., Gür, T. M., Kennedy, B., Macyk, W., Bak, T., Sigmund, W., Yamawaki, M., & Rahman, K. A. (2018). Towards Global Sustainability: Education on Environmentally Clean Energy Technologies. *Renewable and Sustainable Energy Reviews, 81*, 2541–2551.

Öhman, J., & Sund, L. (2021). A Didactic Model of Sustainability Commitment. *Sustainability: Science Practice and Policy, 13*(6), 3083.

Orr, D. W. (1992). *Ecological Literacy: Education and the Transition to a Postmodern World.* State University of New York Press.

Pearson, E., & Degotardi, S. (2009). Education for Sustainable Development in Early Childhood Education: A Global Solution to Local Concerns? *International Journal of Early Childhood, 41*(2), 97.

Phelan, L., McBain, B., Ferguson, A., Brown, P., Brown, V., Hay, I., Horsfield, R., & Taplin, R. (2015). *Learning and Teaching Academic Standards Statement for Environment and Sustainability.* Office for Learning and Teaching.

Pramling Samuelsson, I. (2011). Why We Should Begin Early with ESD: The Role of Early Childhood Education. *International Journal of Early Childhood, 43*(2), 103.

QAA (2021). *Education for Sustainable Development Guidance.* Quality Assurance Agency.

Raworth, K. (2017). *Doughnut Economics: Seven Ways to Think Like a 21st-Century Economist.* Chelsea Green Publishing.

Razak, D. (2018). Decolonising the Paradigm of Sustainable Development through the Traditional Concept of Sejahtera. In Z. Fadeeva, L. Galkute & K. Chhoker (Eds.), *Academia and Communities: Engaging for Change*. United Nations University.

Reid, A. (2019). Climate Change Education and Research: Possibilities and Potentials Versus Problems and Perils? *Environmental Education Research, 25*(6), 767–790.

Renert, M. (2011). Mathematics for Life: Sustainable Mathematics Education. *For the Learning of Mathematics, 31*(1), 20–26.

Rickinson, M., & McKenzie, M. (2021). The Research-Policy Relationship in Environmental and Sustainability Education. *Environmental Education Research, 27*(4), 465–479.

Ryan, A., & Tilbury, D. (2013a). *Flexible Pedagogies: New Pedagogical Ideas*. Higher Education Academy. http://s3.eu-west-2.amazonaws.com/assets.creode.advancehe-document-manager/documents/hea/private/resources/npi_report_1568036616.pdf

Ryan, A., & Tilbury, D. (2013b). Uncharted Waters: Voyages for Education for Sustainable Development in the Higher Education Curriculum. *The Curriculum Journal, 24*(2), 272–294.

Schneider, F., Kallis, G., & Martinez-Alier, J. (2010). Crisis or Opportunity? Economic Degrowth for Social Equity and Ecological Sustainability. Introduction to This Special Issue. *Journal of Cleaner Production, 18*(6), 511–518.

Sjöström, J., & Rydberg, C. (2018). Towards Transdisciplinary Didaktik: Didactic Modelling of Complex Controversial Issues Teaching for Reflexive Bildung and Sustainability. In I. Eilks, S. Markic & B. Ralle (Eds.), *Building Bridges Across Disciplines for Transformative Education and a Sustainable Future* (pp. 3–15). Shaker Verlag.

Smith, C., & Watson, J. (2019). Does the Rise of STEM Education Mean the Demise of Sustainability Education? *Australian Journal of Environmental Education, 35*(1), 1–11.

Stephenson, B., Brody, M., Dillon, J., & Wals, A.E.J. (Ed.). (2013). *International Handbook of Environmental Education Research*. Routledge.

Sterling, S. (2001). *Sustainable Education: Re-visioning Learning and Change*. Schumacher Society.

Sterling, S., Maxey, L., & Luna, H. (2013). In S. Sterling, L. Maxey, & H. Luna (Eds.), *The Sustainable University: Progress and Prospects*. Routledge. https://doi.org/10.4324/9780203101780

Stern, N. (2006). *Stern Review: The Economics of Climate Change*. https://www.osti.gov/etdeweb/biblio/20838308

Stevenson, R. B., Nicholls, J., & Whitehouse, H. (2017). What Is Climate Change Education? *Curriculum Perspectives, 37*(1), 67–71.

Stratford, R., & Wals, A. E. (2020). In Search of Healthy Policy Ecologies for Education in Relation to Sustainability: Beyond Evidence-Based Policy and Post-Truth Politics. *Policy Futures in Education, 18*(8), 976–994.

Sund, L., & Pashby, K. (2020). Delinking Global Issues in Northern Europe Classrooms. *The Journal of Environmental Education, 51*(2), 156–170.

Tilbury. (2013). *Another World Is Desirable: A Global Rebooting of Higher Education for Sustainable Development*. The Sustainable University. https://doi.org/10.4324/9780203101780-15/another-world-desirable-global-rebooting-higher-education-sustainable-development-daniella-tilbury

Tilbury. (2014). Education for Sustainability in Higher Education. *Report Commissioned by UNESCO ESD Secretariat to.* https://www.researchgate.net/profile/Daniella-Tilbury/publication/324828998_Ten_Years_of_Education_for_Sustainability_in_Higher_Education_UNESCO_Commissioned_Report_for_the_Decade_in_Education_for_Sustainable_Development_DESD/links/5ae57ca9458515760ac08826/Ten-Years-of-Education-for-Sustainability-in-Higher-Education-UNESCO-Commissioned-Report-for-the-Decade-in-Education-for-Sustainable-Development-DESD

Tilbury, D. (2011). *Education for Sustainable Development: An Expert Review of Processes And Learning.* Paris: UNESCO.

Tilbury, D. (2012). Learning to Connect: Reflections Along a Personal Journey of Education and Learning for a Sustainable Future in the Context of Rio + 20. *Journal of Education for Sustainable Development, 6*(1), 59–62.

Tilbury, D. (2019). Beyond Snakes and Ladders: Overcoming Obstacles to the Implementation of the SDGs in Higher Education Institutions. In *Implementing the 2030 Agenda at Higher Education Institutions: Challenges and Responses* (pp. 75–81). Global University Network for Innovation (GUNi). http:// www.guninetwork.org/files/guni_publication_-_implementing_the_2030_agenda_at_higher_education_institutions_challenges_and_responses.pdf

Tilbury, D. (2022). *Input Paper: A Whole School Approach to Learning for Environmental Sustainability.* European Union.

Tilbury, D., Alba, D., Mulà, I., Junyent, M., Gutiérrez, J., Serrano, A., & Fonolleda, M. (2019). *Proposal of indicators to embed the sustainable development goals in institutional quality assessment.* Quality Assurance Agency for Higher Education for Andorra (AQUA).

Tilbury, D, and Ryan. Mula, I, A., Doulha, J. Mader, M., Mader, C. (2012). *University Educators for Sustainable Development: Research as Institutional Change.* University of Gloucestershire.

Tilbury, D., Podger, D., & Reid, A. (2004). *Action Research for Change towards Sustainability: Change in Curricula and Graduate Skills towards Sustainability: Final Report.* Australian Government Department of the Environment and Heritage and Macquire University.

Tuck, E., McKenzie, M., & McCoy, K. (2014). Land Education: Indigenous, Post-Colonial, and Decolonizing Perspectives on Place and Environmental Education Research. *Environmental Education Research, 20*(1), 1–23.

UNECE (2023). *Workplan for the Implementation of the United Nations Economic Commission for Europe Strategy for Education for Sustainable Development from 2021 to 2025.* UNECE.

UNEP (2010). *"Here and Now! Education for Sustainable Consumption" Report, published by UNEP.* United Nations Environment Programme.

UNESCO (1975). *Belgrade Charter on Environmental Education.* UNESCO.

UNESCO (2012). *Education for Sustainable Development: Good Practices in Early Childhood.* UNESCO.

UNESCO (2016). *Education for People and Planet: Creating Sustainable Futures for All: Global Education Monitoring Report.* UNESCO.

UNESCO (2020). *Education for Sustainable Development: Partners in Action; Global Action Programme (GAP) Key Partners' Report (2015-2019).* Paris: UNESCO.

van Boeckel, J. (2014). *At the Heart of Art And Earth: An Exploration of Practices in Arts-Based Environmental Education* (pp. 801–802). Aalto University.

Van Poeck, K. (2019). Environmental and Sustainability Education in a Post-Truth Era. An Exploration of Epistemology and Didactics Beyond the Objectivism-Relativism Dualism. *Environmental Education Research, 25*(4), 472–491.

Verlie, B. (2017). Rethinking Climate Education: Climate as Entanglement. *Educational Studies, 53*(6), 560–572.

Wals, A. E. J. (2017). Sustainability by Default: Co-Creating Care and Relationality Through Early Childhood Education. *International Journal of Early Childhood, 49*(2), 155–164.

Wals, A. E. J., Brody, M., Dillon, J., & Stevenson, R. B. (2014). Convergence Between Science and Environmental Education. *Science, 344*(6184), 583–584.

Wessels, K. R., Bakker, C., Wals, A. E. J., & Lengkeek, G. (2022). Rethinking Pedagogy in the Face of Complex Societal Challenges: Helpful Perspectives for Teaching the Entangled Student. *Pedagogy, Culture & Society, 32*(3), 759–776.

7 Education and the Environment in the 2020s

Regenerate and Transition

The Transformation of Education

As we arrive to the 2020s, we find pandemics, war, and economic development shaping global discourses and educational responses to environmental and sustainability concerns. In 2024, conflicts raging in the Middle East, Northern Africa, and Eastern Europe add another layer of complexity and distraction as world leaders wrestle with new regional power dynamics. On the environmental front, the news reports daily on records on droughts, wildfires, ocean temperatures, ice-masses melting, rainfall, and storms. Geopolitical tensions arising over minerals, rare earth metals, water, and land are on the rise as is social unrest and people having to leave their homeland. After the early warning signs presented at the 1972 Stockholm conference, and decades of increasingly strongly worded reports on human-induced climate change, people across the globe are becoming more aware (and worried) about the wellbeing of people and the planet. Increasingly, high-level reports and statements concerning the state of the planet are now also pointing at the critical role of education in transitioning towards a more sustainable world (IPBES, 2019; IPCC, 2022; United Nations, 2024a). Current learning, education, and environmental policies are increasingly being shaped by six key narratives.

First, the imperative to phase out fossil fuels emerges as a central theme, grounded by the agreement formalised in Dubai at the United Nations Framework Convention on Climate Change's (UNFCCC) at COP28, emphasising the necessity of ensuring a safe and sustainable transition (Vatalis et al., 2022). The climate urgency and related loss of biodiversity and increased migration/climate refugees demand a response from education but many schools and school systems do not know how (Irwin, 2020; Rousell & Cutter-Mackenzie-Knowles, 2020). Climate change education is emerging in many countries as a new adjectival education, one that does not necessarily take on an integrative approach.

Second, there is a growing recognition that educating girls plays a pivotal role in addressing environmental concerns, recognising the transformative impact of empowering women in environmental sustainability (see FCDO, 2023). This connects with the gender and wider inclusion agenda

DOI: 10.4324/9781003467007-7

that also calls for a response from all levels of education (United Nations, 2024b).

Third, an emphasis on nature-based solutions underscores the importance of integrating ecological approaches into strategies for sustainable development (Seddon et al., 2020). This re-connects with earlier nature conservation, nature studies, and outdoor education initiatives but this time takes on a more relational and even posthuman turn where decentering the human and providing rights to mature are considered prerequisites for sustainability (Maller, 2021).

Fourth, partnerships are indispensable in financing and scaffolding the transition towards a green future, emphasising collaborative efforts across sectors and stakeholders (Ansell et al., 2022). This narrative connects with social learning but also with the recognition of the importance of boundary crossing and considering multiple perspectives (Chan et al., 2020).

Fifth, a forward-looking perspective underscores the convergence of digital education and environmental education, advocating for a transformative synergy that can foster a more sustainable and environmentally conscious global community. The turn to the digital, artificial intelligence (AI), and videophilia poses both threats to and opportunities for engaging (young) people in sustainability challenges, developing a sense of place and an ethic of care (Edwards & Larson, 2020).

Finally, there is a need for a future-oriented perspective that provides active hope and concrete possibilities for positive change. This is especially important to address rising levels of anxiety about the future, especially among young people (Clayton, 2020). This requires educators, at all levels, to also be engaged, and be competent, in dealing with emotions and engaging in socio-emotional learning (Olsen et al., 2024). This narrative is also underpinned by a sense of fairness and justice, which has consequences for education and generations yet to be born as intergenerational equity issues come to the forefront.

These narratives were featured to varying degrees in the UNESCO report calling for a 'new social contract' that affirms the transformative and empowering potential of education to shape peaceful, just, and sustainable futures. Most significantly, the report formally acknowledged that 'education itself must be transformed' if it is to be able to fulfil this responsibility (UNESCO, 2021c, p. 1). Importantly, the UN Secretary General Antonio Guterres called for the Transformation Education Summit in 2022 (United Nations, n.d.-c), and a commitment to sustainability for current and future generations in the upcoming Summit of the Future 2024 (United Nations, n.d.-b). He established a high-level panel on the teaching profession which made specific deliberations on how best to integrate sustainability into teacher preparation and development (United Nations, 2024b). These developments point to the mainstreaming of education and environment into intergovernmental agendas and elevate education to the top of the global political agenda.

Education as Relevant and Responsive to Current Challenges

What has finally become clear are the limitations to the transmissive, classroom approach to teaching and learning, with the recognition of the critical role of learner engagement and student-centred learning. This is nowhere more apparent than in the ongoing climate change crisis. The conversation in educational circles has continued to move away from individual responsibility and single actions to a level above where politicians, leaders, and heads of schools are all talking and committing to urgent action for the environment (González-Gaudiano & Cartea, 2020; McKenzie et al., 2023). Yet parallel to this a lot of people, including young people themselves, are pointing fingers saying that schools and universities are failing students and young people, as well as our planet, by not seriously engaging with the existential issues of our time (Tilbury, 2023b).

As noted by Kvamme and colleagues (2022), leading figures like Greta Thunberg have reported learning about global warming and the climate emergency in school, highlighting the formal education focus on awareness building. However, school strikes see students leaving the school building and their lessons, in favour of political protests on the street. On the one hand, Huttunen and Albrecht (2021) suggest that the students participating in these strikes promote environmental citizenship. On the other hand, for engaged students to leave the classroom raises difficult questions about the relevance and responsiveness of formal education (Tilbury, 2021). Greta Thunberg and the Fridays-For-Climate movement exemplify how many young people consider that school is failing to give them an adequate understanding of climate change, the environment, and how to live, work, and act more sustainably.

The rise of climate change education corresponds with this movement and can be regarded as another important educational response. At the international policy level this is recognised in SDG target 13.3, which relates climate change to ESD in the need to '*improve education, awareness raising and human and institutional capacity on climate change mitigation, adaptation, impact reduction, and early warning*'. However, while decades of rigorous research have shown that education *about* the environmental and social issues is not enough to create meaningful change (Iyengar & Kwauk, 2021), González-Gaudiano and Meira-Cartea (2022) argue that, instead of focussing on the socio-environmental roots of environmental concerns, climate change education is using the same failed strategy of 'climate literacy', hoping that information on climate change will raise awareness and lead to behavioural change (mostly through science education). The authors argue that this cognitive literacy has contributed to the perpetuation of the climate crisis, as individuals and society are disconnected at the emotional level. Instead, a reorientation of educational processes is needed for learners to think substantially differently. To add to this debate, the examples above on climate strikes illustrate how bottom-up de-institutionalised forms of environmental learning can lead to engagement and action in terms of addressing climate change, which are beginning to receive support from the scientific community in terms of their value and significance (Fisher, 2019; Lotz-Sisitka & Rosenberg, 2022). Interestingly, some authors have argued that the increased public and political interest in climate change have raised environmental concern since its relative decline since the 1992 UN Conference in Rio (González-Gaudiano & Meira-Cartea, 2010).

Besides the UN policy directives that relate to this, the European Union also has responded by launching the Green Deal. The European Council, spurred by the Green Deal, has issued a policy recommendation to member states, highlighting the crucial need for 'education for environmental sustainability' and transforming education so that it includes youth voices on the climate and biodiversity crises. Furthermore, these policies contribute to a shift towards whole-institution approaches to environmental learning that continue to break down silos and extend efforts beyond champion teachers (European Council, 2022). Underpinning this document, and its accompanying handbook (European Commission, 2022), is a recognition that teacher development and training for sustainability must become a priority across the European Union if this vision is to be realised. Alongside the continued impetus to work towards the UN Agenda 2030, this bold move has the possibility to dramatically change the policy environment across Europe with significant incentives provided for the change via financial support and diverse collaborative platforms as well as mobility programmes.

The special emphasis which the above report places on the environmental pillar can be witnessed in what we can term 'the (re)turn to the ecological' in education and learning. There is a strong push towards the redesigning of 'sustainable smart' cities (Ahad et al., 2020), the greening of cities in response to the crises of climate change (Bayulken et al., 2021), as well as the greening of school yards and buildings (van Dijk-Wesselius et al., 2018). In response to the growing pressures of urbanisation and technophilia (at the expense of biophilia), there is also a renewed push for outdoor education and more relational forms of place-based education (Gilbertson et al., 2022). This includes wild pedagogies (Blenkinsop et al., 2020) but also living labs, which are intentional spaces for innovation, experimentation, and boundary crossing in sustainability education (Macintyre et al., 2019; von Wirth et al., 2019).

Internationally, the Berlin declaration on Education for Sustainable Development underscored the pivotal role of transformative learning in fostering positive outcomes for both individuals and the planet (UNESCO, 2021a). Concurrently, the United Nations Framework Convention on Climate Change's Action for Climate Empowerment (ACE) initiative accentuated the significance of climate action (United Nations, n.d.-a). Although the recent 28th Conference of the Parties (COP 28), under the joint leadership of the UK Department for Education and UNESCO, constituted a crucial milestone by recognising education as an enduring solution to the climate crisis, a degree of ambiguity persists regarding the specific nature and structure that such educational endeavours should adopt.

Finally, it is important to remember that a key characteristic of emerging educational responses is the move beyond purely cognitive ways of knowing. As demonstrated by the climate strikes, the fees-must-fall movement in South Africa, and the yellow vests protests in France, there is a great amount of frustration, anger, and ecoanxiety (Ojala, 2021; Ojala et al., 2021). This is expressed by people going to the streets to protest, as it is considered the only way to make an impact. Noting this frustration, some sustainability scholars begin to flag the importance of disruptive capacity building, counter-hegemonic thinking and acting, and transgressive learning as more radical and activist forms of education (Pedersen et al., 2022; Wals, 2022).

It would appear that developments in paradigms and approaches noted in the previous decades have never truly translated into mainstream practice. While educational systems are helping students understand and become aware of the urgent action needed – much like the focus of Environmental Education in the 1970s –these approaches have proved ineffective in providing the capabilities that help learners make the change. Instead, this lack of empowerment has resulted in the rapidly growing phenomenon of eco-anxiety amongst individuals (Ojala et al., 2021; Pihkala, 2020). This highlights once again the point made by environmental educator David Orr (1994) that education must benefit the planet as well as the learner, orientating education towards sustainability worldwide, for both young and old, in ways that make sense to the contexts and challenges that citizens face.

New Transgressive and Regenerative Strands of Education

A pertinent branch of emerging education that recognises the above shortcomings is that of 'regenerative education', which has been described as the next wave of sustainability (Gibbons, 2020). Transgressing the goals of sustainability, regenerative education aims for 'living systems in which whole-system health and wellbeing increase continually' (Gibbons, 2020, p. 1). Based on a holistic worldview and paradigm, regenerative sustainability integrates science and practice, different ways of knowing, and inner and outer dimensions of sustainability necessary for systemic transformation. On the one hand, regenerative education involves empathic qualities of healing and restoration, accepting the damage of human intervention on planet earth, and working on how to heal this damage (Mehmood et al., 2020). As Reed (2007) notes, we can best engage in healing in the places we inhabit, in the communities we live. In this sense, place-based learning processes and research carried out in community contexts assume a key role in this perspective of regenerative education. On the other hand, regenerative education also involves disruptive elements of breaking with the status quo, addressing power structures and norms which act as barriers to bringing about more regenerative futures. The aforementioned characteristics of regenerative education strongly resonate with strong counter-hegemonic traditions (Escobar, 2020), both in terms of education and development, as well as indigenous worldviews based on a non-binary understandings of mind and body, and humanity and nature, providing interesting sites to refine this emergent conception of education. The concept of regeneration is embryonic in UNESCO's educational approaches, being introduced into documents such as the Futures of Education initiative (UNESCO, 2021b).

An important aspect of such regenerative and transgressive forms of learning are counter-hegemonic encounters that identify and uproot systems of oppression and marginalisation (Lotz-Sisitka et al., 2015; Macintyre et al., 2018). Regenerative education approaches to addressing climate change can involve arts-based and participatory methods (Bentz & O'Brien, 2019; Macintyre et al., 2019), as well as transgressing the boundaries between higher education and community-based learning (Macintyre et al., 2020; van den Berg et al., 2022). The rapid rise of LivingLabs and ChallengeLabs as learning spaces co-created by different stakeholders in real physical spaces focusing on wicked sustainability challenges as they arise in a neighbourhood, community, city, or even region can be seen as concrete manifestations of boundary crossing, blended forms of learning and linking science and society (Van der Wee-Bedeker et al., 2024). Citizen science, sometimes as a part of LivingLabs, also has become in vogue over the last few years (Sauermann et al., 2020).

Youth and Marginalised Voices in Education and the Climate Crisis

An issue that has come through strongly in recent debates, such as on the climate strikes, is that the learner and marginalised peoples have little voice in terms of bringing about educational changes. While participation became a buzzword in the 2000s (Alejandro Leal, 2007), children and young learners who are the main beneficiaries of our education system have been underrepresented in the design of the educational process and the curriculum, despite representing a powerful force for social change (Bentz & O'Brien, 2019). We see an interest again in futures thinking, which was around in the 1970s and 1980s as a smaller strand of work, but gains momentum as intergenerational justice becomes a major issue within the climate crisis. We can see this in the 2023 IPCC synthesis report that warns how future generations will be disproportionately affected by climate change (IPCC, 2023). Ironically, it is the younger generations and the ones still to come who are the most vulnerable to changes in climate, although they are the least responsible for this crisis. This has been referred to as the double injustice of climate change (Füssel, 2010).

Youth respond to the worsening climate crisis, and feeling of disempowered by mainstream processes in different ways. Some young people simply disconnect from social engagement, as seen by the increasing numbers of young people intensely engaged in virtual reality gaming and other screen time activities. Significantly, researchers point to how engagement levels are lowest in youth from socio-cultural backgrounds that differ from that of the core cultural group of a school, town, or city (Tilbury, 2023a). As mentioned in the section above, the other response by youth is to engage in protests or disruptive actions, whereby young people are reclaiming agency in building their own futures. This is shown by school strikes, legal challenges to governments, and online and offline climate activism. There is therefore a clear role for education to provide ways to conceptualise futures – to recreate, transgress, and transform imperfect presents – by engaging learners in defining and meeting the needs of a future that they are reimagining (Corcoran et al., 2017).

Fueled by the increased urgency, many NGOs, government bodies, and scientists have turned to education in trying to influence what is being taught in schools. This leads to a revival of the tensions noted in the 1980s and 1990s between instrumental and emancipatory interpretations of environmental and sustainability education. As Trott (2021) notes, while the shear massive scale of climate challenges may suggest that only structural and policy top-down changes will make a sufficient impact on lowering carbon emissions, the increasing everyday climate activism of children and youth, and its substantial transformative potential, represents another means to assess progress towards a more sustainable future (Tilbury, 2023a; see Figure 7.1). Some scholars are critical of climate change education as it might become overly instrumental and become another adjectival education. They argue for a more emancipatory and integrated approach that is built into education, rather than added on (Nusche et al., 2024; Reid, 2019). In this way, the climate crisis can be seen as a symptom of a much deeper crisis in Western civilisation and not as a problem that can be solved (Luzzatto, 2022).

The previous decade centred on the notion of young children as agents of change for sustainability who are able to engage with complex sustainability problems and can creatively work on solutions. Underneath is an assumption that children can be viewed as agents who can be taught to be ethical and rational to care for and safeguard the world. In the 2020s, some scholars critique this perspective as being inherently anthropocentric in that it neglects the agentic characteristics of the non-human world and thereby creating an artificial separation between the human child and the wider more-than-human world (Malone et al., 2020; see Somerville, 2020; Weldemariam & Wals, 2020). Instead, they plead for abandoning the idea of the rational, ethical, and

---

The following are examples capture the range of student engagement and participation activities in sustainability from Tilbury (2023a):

- Young people being **consulted and involved** in plans for the upgrading of their school playground or local park.
- Young people **inquiring into issues** related to food served in the school canteen and forming an opinion or position that informs personal choices.
- Young people **calling for** greater youth representation in committees addressing climate change.
- Young people **taking part** and actively contributing to an event that raises awareness about new plans for sustainable housing in their community.
- Young people **joining youth committees or groups** to protect local biodiversity and habitats.
- Young people being **given responsibilities to look after** green areas in school
- Young people **leading** a school magazine where they **become reporters and critical commentators on key sustainability issues** relevant to young people.
- Young people **lobbying** a local supermarket to reduce its dependence on plastics and offer alternatives to consumers.
- Young people being a key stakeholder in **school decision-making bodies** that shape how the school practises sustainability.

Figure 7.1 Student engagement and participation in sustainability

agentic child, in favour of the idea of the unfolding relational child, and its implications for sustainability. The question is now becoming: how to support teachers pedagogically in their efforts to support the development of childrens' agency while inviting them to live their entanglement in the world (Borg & Samuelsson, 2022; Weldemariam & Wals, 2020). As classrooms and curriculum become policy tools for leaders combating climate change (Bond, 2023), it is important that young people have an opportunity to shape and drive learning experiences, combating the return of instrumentalist influences on Environmental Education.

Intergenerational Justice and Inclusivity

Climate change is an inescapably intergenerational issue: its detrimental impacts, driven by unsustainable human activity, pose threats to justice and equity between current and future generations. The climate emergency and ascendancy of climate agendas in education have meant that intergenerational justice has become one of the key narratives for the 2020s. While the Brundtland definition of sustainability signalled a future focus of a 'form of development that meets the needs of the present without compromising the ability of future generations to meet their own needs' (WCED, 1987), the section above highlights the frustration of younger generations to address pressing environmental challenges inherited from their predecessors.

Essentially, the intergenerational justice agenda is focused on halting short-termism in political and economic decision-making. It has gained traction in a world where it is increasingly clear that many leaders are focused on immediate concerns, short-term gains, and electoral cycles which can result in the exploitation of ecosystems and natural environments. Girardet (2023) argues that this kind of thinking invariably leads to compromised values and ethics, affecting the long-term prospects of humanity with education failing to question who benefits most from the status quo.

Moreover, access to quality education is often unequally distributed, exacerbating intergenerational disparities, and making learning for sustainability a more significant challenge in some parts of the world. Communities with fewer educational resources may face challenges in equipping younger generations with the knowledge and skills needed to navigate and address environmental issues effectively. Addressing this, the Secretary General's Summit for the Future asks leaders to develop climate change education strategies that 'leave no one behind' and are inclusive of vulnerable young people as well as communities that face social inequalities. Although still marginal in influence, some teachers have created justice-centred learning experiences, where students are able to consider the impact of systemic racism on the biodiversity of plants and animals in different cities through a pollution and climate change lens as well as apply what they learned to unpacking developments and disparities in their local communities (Bond, 2023).

The Twin Transitions: Digital and Green

In schools where pupils have access to digital technologies, new opportunities for environmental learning are emerging. This can be seen in Europe, where explicit efforts to twin the green and digital transitions are supporting learners to challenge the way we presently live, learn, and work as well as consider the future (Muench et al., 2022). Significantly, the digital transition leverages technology to enhance learning experiences, providing interactive tools, virtual simulations, and global connectivity for a comprehensive and experiential understanding of environmental challenges. The synthesis of these transitions can offer a holistic approach, empowering learners with the tools to navigate a technologically advanced world, while cultivating an environmentally conscious ethos for a sustainable future. Indeed, pathways leading to the green and digital transitions are multifaceted and often interconnected, whereby the European Commission has underlined that an environmentally sustainable, circular, and climate-neutral economy cannot be attained without harnessing the new technologies.

At the same time, there are also concerns of the highly addictive digitally mediated experience taking over the embodied lived experience and AI and algorithms taking over our natural intelligence and our own capacity for deep thinking (Carr, 2020; Farrokhnia et al., 2023; Sutton, 2020). Within this green transition, an important aspect is the increasing role of social media in how we perceive and engage with sustainability. Learning experiences have shifted to become more hybrid, with a generational sustainability survey carried out in 2023 (EYGM, 2023), showing that social media is the most common, but least trusted source for sustainability education. A key question is how to balance and connect the digital world with the natural world in a way that might help people connect with each other and with nature, thereby potentially accelerating sustainability as opposed to unsustainability (Edwards & Larson, 2020).

One way this is being addressed in the 2020s is through competency-based frameworks. For example, the DigiComp provides a digital competence framework aimed at enhancing the skills of citizens, assisting policymakers in crafting policies that promote digital literacy, and strategising educational and training programs to enhance the digital proficiency of targeted demographics (Vuorikari et al., 2022). Another example is GreenComp, which was developed as a European sustainability competency framework, and presented as one of the policy measures in the European Green Deal, aiming to stimulate education on environmental sustainability within the European Union (Bianchi et al., 2022). Both agendas are supportive of whole-school approaches, participatory and engaged learning, of extending learning experiences beyond the classroom and, thus, ultimately bringing educational innovation into schools (Tilbury, 2024). Whilst digital technologies excite many educationists seeking a renewal of educational purposes and praxis, it also worries environmentalists as the carbon footprint of these technologies can exceed that of the aviation industry.

The Pedagogy of Transition

The question of how transition education systems towards sustainability has remained at the forefront of environmental concerns. Scholars continued to unpack theoretical frameworks and the ideological underpinnings of sustainability education efforts with some pointing to how the Sustainable Development Goals (SDGs) are instrumentalised and deepen marketisation in the present world economy (Bonn, 2021). The Times Higher Education Impact Rankings, which made their mark in the early years of the decade, dominated much of the higher education sector's interest in the SDGs as Universities sought a competitive advantage through the impact agenda.

A contrasting approach was adopted by the Global University Network for Innovation (GUNI), a UN affiliated organisation, that had previously generated significant and cutting-edge literature on how to redesign higher education institutions for sustainability. In December 2023 it launched its 'International Call for Action' and forged a new pathway for its membership. The programme sought to drive the transformation of higher education institutions through a series of institutional change projects that built capability in teams and created transnational spaces for training, reflection, and exchange (GUNI, 2023). After co-hosting the World Congress in Higher Education, GUNI had understood that the sector needed a new pedagogy for transition.

The need for new pedagogical theories of change also struck a chord with education policymakers who were seeking to advance the agenda in schools (see UK Climate and Sustainability Education Strategy (Department for Education, 2023)). Whilst teachers embraced, at least in principle, their responsibilities in this area, a global study of educators found that one in four teachers did not feel confident or competent in creating learning opportunities for sustainability (UNESCO, 2021d). Teacher preparation regained importance during this decade with the European Education Network for Economics Education commissioning a study in 2023, which identified catalytic pedagogies and strategies for change in teacher education (Mulà & Tilbury, 2023). Also on the rise was interest in the professional standards of teachers with Scotland, for example, formally adopting teacher professional frameworks that would support career progression for teachers in learning for sustainability (GTCS, 2021).

Julie Davis' and Sue Elliotts' 'Young Children and the Environment' also articulated a pedagogy of transition but for early childhood educators (Davis & Elliott, 2023). Noting that educational horizons are changing rapidly and dominated by a back-to-basics approach to learning, the book provides the rational but also tangible strategies for effecting change in early childhood learning. It invites educators and children to grow food together in their learning centres, establish gardens in the backyards, consider how green the fridge or supermarket offerings are, and engage indigenous elders and seniors from the community as the focal point for early childhood learning. The text shows how these pedagogies can join together, and in ways that reposition Environmental Education and the schooling experience.

Building upon the work of Pramling-Samuelson, Davis and Elliott (2023) drew together thinking around the pedagogy of play and recognised that playful pedagogies give young pupils the flexibility to make sense of the world and engage in personally meaningful activities that drive engagement with the environment. Arriving at similar conclusions but seeking to engage adolescents and adults in climate and biodiversity issues, UNEP convened the 'Playing for the Planet' Alliance. Noting the significant fact that the gaming industry reaches 1 in 3 people on the planet, UNEP is encouraging the sector to look at pedagogies that can inspire young people to act in support of the environment and thus extend the influence of Environmental Education, whilst at the same time encouraging the gaming industry to enact new measures and standards for decarbonisation (UNEP, 2024). See Figure 7.2 for a summary of education and the environment in the 2020s.

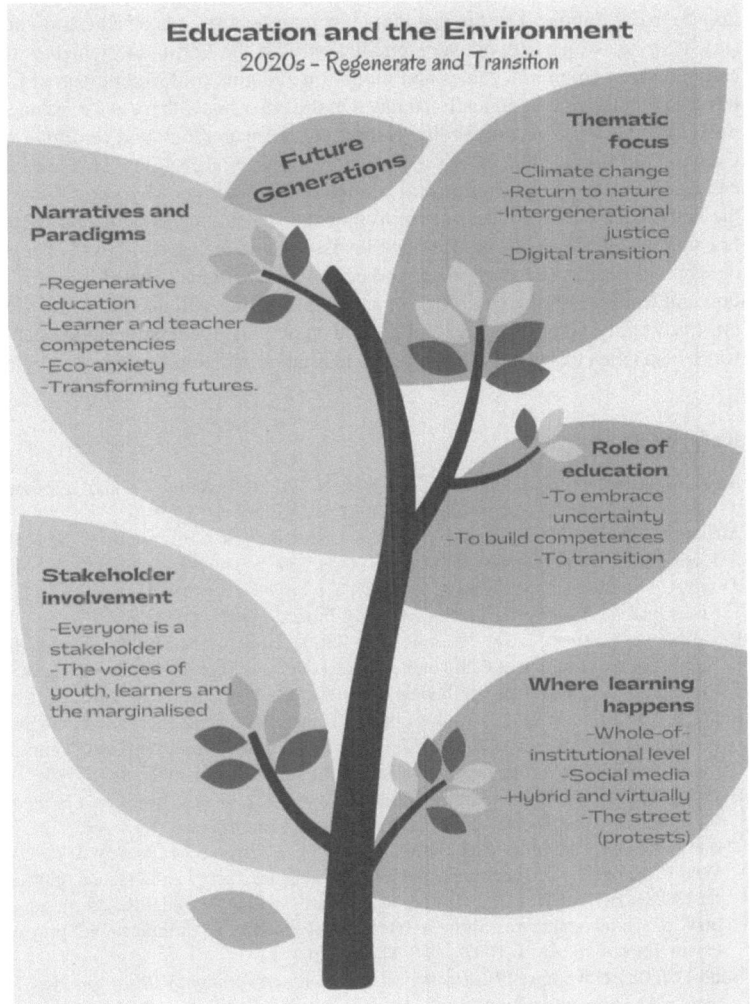

Figure 7.2 Summary of education and the environment in the 2020s

Moving from the 2020s into the Future

As we are approaching the mid-point of the current decade, it is becoming clear that the decade will be defining for the future of the Earth. Agenda 2030 has set clear goals with which to work towards, and education is a vital component in this endeavour. It is, however, difficult to gauge how education and learning for the environment and sustainability will continue into this decade and the future. On the one hand, exciting new strands of education are (re)emerging, which propose a reconciliation with the Earth, and which critically confront colonial legacies and modern paradigms of development which have shaped the role of education today. On the other hand, there is a continued push for efficiency, accountability, testing and measurement, and returning to the good old 'basics' that clearly works in the opposite directions. This tension can lead to immense frustration and eco-anxiety of learners who seem to have the knowledge and awareness of environmental issues such as climate change, but not the ability to address such complex issues, and the capabilities to change deeply ingrained systems of power and control. Pedagogically speaking education and learning for the environment and sustainability will need to work with this uncertainty, complexity, ambiguity, and anxiety in regenerative and hopeful ways, also when this means encountering resistance and going against the grain.

References

Ahad, M. A., Paiva, S., Tripathi, G., & Feroz, N. (2020). Enabling Technologies and Sustainable Smart Cities. *Sustainable Cities and Society*, *61*, 102301.

Alejandro Leal, P. (2007). Participation: the Ascendancy of a Buzzword in the Neo-Liberal Era. *Development in Practice*, *17*(4–5), 539–548.

Ansell, C., Sørensen, E., & Torfing, J. (2022). *Co-creation for Sustainability: The UN SDGs and the Power of Local Partnerships*. Emerald Publishing.

Bayulken, B., Huisingh, D., & Fisher, P. M. J. (2021). How Are Nature Based Solutions Helping in the Greening of Cities in the Context of Crises Such as Climate Change and Pandemics? A Comprehensive Review. *Journal of Cleaner Production*, *288*, 125569.

Bentz, J., & O'Brien, K. (2019). Art for Change: Transformative Learning and Youth Empowerment in a Changing Climate. *Elementa: Science of the Anthropocene*, *7*. https://online.ucpress.edu/elementa/article-abstract/doi/10.1525/elementa.390/112520

Bianchi, G., Pisiotis, U., & Cabrera Giraldez, M. (2022). *GreenComp The European Sustainability Competence Framework*. European Commission.

The full reference should be: Blenkinsop, S., Jickling, B., Morse, M., Jensen, A. (2020). Wild Pedagogies: Six Touchstones for Childhoodnature Theory and Practice. In: Cutter-Mackenzie-Knowles, A., Malone, K., Barratt Hacking, E. (eds) Research Handbook on Childhoodnature . Springer International Handbooks of Education. Springer, Cham. https://doi.org/10.1007/978-3-319-67286-1_32

Bond, H. (2023, November 29). *Advancing Climate Action Through Justice-Centered Climate Education*. International Institute for Sustainable Development. https://sdg.iisd.org/commentary/guest-articles/advancing-climate-action-through-justice-centered-climate-education/

Bonn, J. (2021, May). *Justice in the Sustainability Curriculum.* Great Transition Initiative. https://greattransition.org/gti-forum/pedagogy-transition-juego.

Borg, F., & Samuelsson, I. P. (2022). Preschool Children's Agency in Education for Sustainability: The Case of Sweden. *European Early Childhood Education Research Journal, 30*(1), 147–163.

Carr, N (2020) *The Shallows: What the Internet Is Doing to Our Brains.* W. W. Norton & Company.

Chan, K. M. A., Boyd, D. R., Gould, R. K., Jetzkowitz, J., Liu, J., Muraca, B., Naidoo, R., Olmsted, P., Satterfield, T., Selomane, O., Singh, G. G., Sumaila, R., Ngo, H. T., Boedhihartono, A. K., Agard, J., de Aguiar, A. P. D., Armenteras, D., Balint, L., Barrington-Leigh, C., & Brondízio, E. S. (2020). Levers and Leverage Points for Pathways to Sustainability. *People and Nature, 2*(3), 693–717.

Clayton, S. (2020). Climate Anxiety: Psychological Responses to Climate Change. *Journal of Anxiety Disorders, 74,* 102263.

Corcoran, P. B., Weakland, J. P., & Wals, A. E. J. (2017). *Envisioning Futures for Environmental and Sustainability Education.* Wageningen Academic Publishers.

Davis, J., & Elliott, S. (2023). *Young Children and the Environment: Early Education for Sustainability.* Cambridge University Press.

Department for Education (2023, December 20). *Sustainability and Climate Change: A Strategy for the Education and Children's Services Systems.* GOV.UK. https://www.gov.uk/government/publications/sustainability-and-climate-change-strategy/sustainability-and-climate-change-a-strategy-for-the-education-and-childrens-services-systems

Edwards, R. C., & Larson, B. M. H. (2020). When Screens Replace Backyards: Strategies to Connect Digital-media-Oriented Young People to Nature. *Environmental Education Research, 26*(7), 950–968.

Escobar, A. (2020). *Pluriversal Politics: The Real and the Possible.* Duke University Press.

European Commission (2022). *Commission Staff Working Document: Accompanying The Document Proposal for a Council Recommendation on Learning for Environmental Sustainability.* European Commission.

European Council (2022). *Proposal for a Council Recommendation on Learning for Environmental Sustainability – Adoption.* The Council of the European Union. https://data.consilium.europa.eu/doc/document/ST-9242-2022-INIT/en/pdf

EYGM (2023). *Generational Sustainability Survey 2023.* https://assets.ey.com/content/dam/ey-sites/ey-com/en_gl/topics/corporate-responsibility/ey-ja-2023-sustainability-report-27-july-2023.pdf

Farrokhnia, M., Banihashem, S. K., Noroozi, O., & Wals, A. (2023). A SWOT Analysis of ChatGPT: Implications for Educational Practice and Research. *Innovations in Education and Teaching International, 61*(3), 460–474.

FCDO (2023). *Addressing the Climate, Environment, and Biodiversity Crises in and Through Girls' Education.* Foreign and Commonwealth Development Office, UK Government.

Fisher, D. R. (2019). The Broader Importance of# FridaysForFuture. *Nature Climate Change, 9,* 430–431.

Füssel, H.-M. (2010). How Inequitable Is the Global Distribution of Responsibility, Capability, and Vulnerability to Climate Change: A Comprehensive Indicator-Based Assessment. *Global Environmental Change: Human and Policy Dimensions, 20*(4), 597–611.

Gaudiano, & Cartea (2020). Educación para el cambio climático:¿ educar sobre el clima o para el cambio? *Perfiles Educativos*. https://perfileseducativos.unam.mx/iisue_pe/index.php/perfiles/article/view/59464

Gibbons, L. V. (2020). Regenerative—The New Sustainable? *Sustainability*, *12*(13), 5483. https://doi.org/10.3390/su12135483

Gilbertson, K., Ewert, A., Siklander, P., & Bates, T. (2022). *Outdoor Education: Methods and Strategies*. Human Kinetics.

Girardet, H. (2023). The Future – What Future? *The Ecologist*.

González-Gaudiano, E., & Meira-Cartea, P. (2010). Climate Change Education and Communication: A Critical Perspective on Obstacles and Resistances. In F. Kagawa & D. Selby (Eds.), *Education and Climate Change*. Routledge.

González-Gaudiano, E., & Meira-Cartea, P. A. (2022). The Banality of Climate Collapse: What Can Education Do? González Gaudiano. In H. Lotz-Sisitka & E. Rosenberg (Eds.), *NORRAG: Education in Times of Climate Change* (pp. 19–22).

GTCS (2021). *Overview of Learning for Sustainability in Professional Standards*. The General Teaching Council for Scotland. https://www.gtcs.org.uk/wp-content/uploads/2021/10/overview-learning-forsustainability-professional-standards.pdf

GUNI (2023). *International Call for Action 2023-2026*. GUNI ICA.

Huttunen, J., & Albrecht, E. (2021). The Framing of Environmental Citizenship and Youth Participation in the Fridays for Future Movement in Finland. *Fennia*, *199*(1). https://doi.org/10.11143/fennia.102480

IPBES (2019). In E. S. Brondizio, S. D. J. Settele & H. T. Ngo (Eds.) *Global Assessment Report on Biodiversity and Ecosystem Services of the Intergovernmental Science-Policy Platform on Biodiversity and Ecosystem Services*. IPBES Secretariat.

IPCC (2023). IPCC, 2023: Summary for Policymakers. In IPCC (Ed.), *Climate Change 2023: Synthesis Report.Contribution of Working Groups I, II and III to the Sixth Assessment Report of the Intergovernmental Panel on Climate Change* (pp. 1–34). IPCC.

IPCC (2022). Climate Change 2022: Impacts, Adaptation, and Vulnerability. In H.-O. Pörtner, D.C. Roberts, M. Tignor, E.S. Poloczanska, K. Mintenbeck, A. Alegría, M. Craig, S. Langsdorf, S. Löschke, V. Möller, A. Okem, & B. Rama (Eds.). *Contribution of Working Group II to the Sixth Assessment Report of the Intergovernmental Panel on Climate Change*. Cambridge University Press.

Irwin, R. (2020). Climate Change and Education. *Educational Philosophy and Theory*, *52*(5), 492–507.

Iyengar, R., & Kwauk, C. T. (2021). *Curriculum and Learning for Climate Action: Toward an SDG 4.7 Roadmap for Systems Change*. Brill.

Kvamme, O., Sinnes, A., & Wals, A. (2022). School Strikes as Catalysts for Rethinking Educational Institutions, Purposes and Practices. In H. Lotz-Sisitka & E. Rosenberg (Eds.), *NORRAG: Education in Times of Climate Change* (pp. 51–54).

Lotz-Sisitka, H., & Rosenberg, E. (Eds.). (2022). *NORRAG: Education in Times of Climate Change*. NORRAG.

Lotz-Sisitka, H., Wals, A. E. J., Kronlid, D., & McGarry, D. (2015). Transformative, Transgressive Social Learning: Rethinking Higher Education Pedagogy in Times of Systemic Global Dysfunction. *Current Opinion in Environmental Sustainability*, *16*, 73–80.

Luzzatto, L. E. (2022). *Intergenerational Challenges and Climate Justice: Setting the Scope of Our Obligations*. Routledge.

Macintyre, T., Chaves, M., Monroy, T., Zethelius, M. O., Villarreal, T., Tassone, V. C., & Wals, A. E. J. (2020). Transgressing Boundaries between Community Learning and Higher Education: Levers and Barriers. *Sustainability: Science Practice and Policy*, *12*(7), 2601.

Macintyre, T., Lotz-Sisitka, H., Wals, A., Vogel, C., & Tassone, V. (2018). Towards Transformative Social Learning on the Path to 1.5 Degrees. *Current Opinion in Environmental Sustainability*, *31*, 80–87.

Macintyre, T., Monroy, T., Coral, D., Zethelius, M., Tassone, V., & Wals, A. E. J. (2019). T-Labs and Climate Change Narratives: Co-Researcher Qualities in Transgressive Action-Research. *Action Research*, *17*(1), 63–86.

Maller, C. (2021). Re-Orienting Nature-Based Solutions With More-than-Human Thinking. *Cities*, *113*, 103155.

Malone, K., Tesar, M., & Arndt, S. (2020). *Theorising Posthuman Childhood Studies*. Springer Singapore.

McKenzie, M., Henderson, J., & Nxumalo, F. (2023). Climate Change and Educational Research: Mapping Resistances and Futurities. *Research in Education*, *117*(1), 3–10.

Mehmood, A., Marsden, T., Taherzadeh, A., Axinte, L. F., & Rebelo, C. (2020). Transformative Roles of People and Places: Learning, Experiencing, and Regenerative Action Through Social Innovation. *Sustainability Science*, *15*(2), 455–466.

Muench, S., Stoermer, E., Jensen, K., Asikainen, T., Salvi, M., & Scapolo, F. (2022). *Towards a Green and Digital Future*. Publications Office of the European Union.

Mulà, I., & Tilbury, D. (2023). *Teacher Education for the Green Transition and Sustainable Development*. Publications Office of the European Union.

Nusche, D., Rabella, M. F., & Lauterbach, S. (2024). *Rethinking Education in the Context of Climate Change: Leverage Points for Transformative Change*. OECD. https://www.oecd-ilibrary.org/education/rethinking-education-in-the-context-of-climate-change_f14c8a81-en

Ojala, M. (2021). To Trust or Not to Trust? Young people's Trust in Climate Change Science and Implications for Climate Change Engagement. *Children's Geographies*, *19*(3), 284–290.

Ojala, M., Cunsolo, A., Ogunbode, C. A., & Middleton, J. (2021). Anxiety, Worry, and Grief in a Time of Environmental and Climate Crisis: A Narrative Review. *Annual Review of Environment and Resources*, *46*(1), 35–58.

Olsen, E. K., Lawson, D. F., McClain, L. R., & Plummer, J. D. (2024). Heads, Hearts, and Hands: A Systematic Review of Empirical Studies About eco/climate Anxiety and Environmental Education. *Environmental Education Research*, 1–28.

Orr, D. W. (1994). *Earth in Mind: On Education, Environment, and the Human Prospect*. Island Press.

Pedersen, H., Windsor, S., Knutsson, B., Sanders, D., Wals, A., & Franck, O. (2022). Education for Sustainable Development in the "Capitalocene." *Educational Philosophy and Theory*, *54*(3), 224–227.

Pihkala, P. (2020). Eco-Anxiety and Environmental Education. *Sustainability: Science Practice and Policy*, *12*(23), 10149.

Reed, B. (2007). Shifting from "sustainability" to Regeneration. *Building Research and Information*, *35*(6), 674–680.

Reid, A. (2019). Climate Change Education and Research: Possibilities and Potentials Versus Problems and Perils? *Environmental Education Research*, *25*(6), 767–790.

Rousell, D., & Cutter-Mackenzie-Knowles, A. (2020). A Systematic Review of Climate Change Education: Giving Children and Young People a "voice" and a "hand" in Redressing Climate Change. *Children's Geographies, 18*(2), 191–208.

Sauermann, H., Vohland, K., Antoniou, V., Balázs, B., Göbel, C., Karatzas, K., Mooney, P., Perelló, J., Ponti, M., Samson, R., & Winter, S. (2020). Citizen Science and Sustainability Transitions. *Research Policy, 49*(5), 103978.

Seddon, N., Chausson, A., Berry, P., Girardin, C. A. J., Smith, A., & Turner, B. (2020). Understanding the Value and Limits of Nature-Based Solutions to Climate Change and Other Global Challenges. *Philosophical Transactions of the Royal Society of London. Series B, Biological Sciences, 375*(1794), 20190120.

Somerville, M. (2020). Posthuman Theory and Practice in Early Years Learning. In A. Cutter-Mackenzie-Knowles, K. Malone & E. Barratt Hacking (Eds.), *Research Handbook on Childhoodnature: Assemblages of Childhood and Nature Research* (pp. 103–127). Springer International Publishing.

Sutton, T. (2020). Digital Harm and Addiction: An Anthropological View. *Anthropology Today, 36*(1), 17–22.

Tilbury, D. (2021). *Honoris Causa.* University of Girona.

Tilbury, D. (2024). *The Twin Transitions: Digital and Sustainability Learning in Schools Working Group on Schools: Learning for Sustainability.* European Commission.

Tilbury, D. (2023a). *Student Voices, Engagement and Action in Learning for Sustainability Working Group on Schools: Learning for Sustainability Input Paper.* Brussels: European Commission.

Tilbury, D. (2023b). *Youth Engagement in Schools: Student Voices, Participation, and Action in Learning for Sustainability. European Education Area Strategic Framework, Working Group on Schools: Learning for Sustainability.* European Commission.

Trott, C. D. (2021). What Difference Does It Make? Exploring the Transformative Potential of Everyday Climate Crisis Activism by Children and Youth. *Children's Geographies, 19*(3), 300–308.

UNEP (2024). *Playing for the Planet.* UNEP. https://www.unep.org/explore-topics/education-environment/what-we-do/playing-planet

UNESCO (2021a). *Berlin Declaration on Education for Sustainable Development.* UNESCO. https://en.unesco.org/sites/default/files/esdfor2030-berlin-declaration-en.pdf

UNESCO (2021b). *International Commission on the Futures of Education Progress Update.* UNESCO.

UNESCO (2021c). *Reimagining our Futures Together – A New Social Contract for Education.* UNESCO.

UNESCO (2021d). *Teachers Have Their Say: Motivation, Skills and Opportunities to Teach Education for Sustainable Development and Global Citizenship.* UNESCO. https://unesdoc.unesco.org/ark:/48223/pf0000379914

United Nations (n.d.-a). *Action for Climate Empowerment.* United Nations Climate Change. Retrieved March 2024, from https://unfccc.int/topics/education-and-youth/big-picture/ACE

United Nations (n.d.-b). *Summit of the Future.* United Nations. Retrieved January 2024, from https://www.un.org/en/summit-of-the-future

United Nations (n.d.-c). *Transforming education summit.* United Nations. Retrieved January 2024, from https://www.un.org/en/transforming-education-summit/about

United Nations (2024a). *Transforming the Teaching Profession: Recommendations and Summary of Deliberations of the United Nations Secretary-General's High Level Panel on the Teaching Profession.* International Labour Office.

United Nations (2024b). *Transforming the Teaching Profession: Recommendations and Summary of Deliberations of the United Nations Secretary General's High-Level Panel on the Teaching Profession.* International Labour Office.

van den Berg, B., Poldner, K., Sjoer, E., & Wals, A. (2022). Practises, Drivers and Barriers of an Emerging Regenerative Higher Education in The Netherlands—A Podcast-Based Inquiry. *Sustainability: Science Practice and Policy, 14*(15), 9138.

Van der Wee-Bedeker, M., Tassone, V., Wals, A. E. J., & Toxler, P. (2024). Characteristics and Challenges of Teaching and Learning in Sustainability-Oriented Living Labs within Higher Education: A Literature Review. *International Journal of Sustainability in Higher Education, 25*(9), 255–277.

van Dijk-Wesselius, J. E., Maas, J., Hovinga, D., van Vugt, M., & van den Berg, A. E. (2018). The Impact of Greening Schoolyards on the Appreciation, and Physical, Cognitive and Social-Emotional Well-Being of Schoolchildren: A Prospective Intervention Study. *Landscape and Urban Planning, 180,* 15–26.

Vatalis, K. I., Avlogiaris, G., & Tsalis, T. A. (2022). Just Transition Pathways of Energy Decarbonization Under the Global Environmental Changes. *Journal of Environmental Management, 309,* 114713.

von Wirth, T., Fuenfschilling, L., Frantzeskaki, N., & Coenen, L. (2019). Impacts of Urban Living Labs on Sustainability Transitions: Mechanisms and Strategies for Systemic Change Through Experimentation. *European Planning Studies, 27*(2), 229–257.

Vuorikari, R., Kluzer, S., & Punie, Y. (2022). *DigComp 2.2: The Digital Competence Framework for Citizens – With New Examples of Knowledge, Skills and Attitudes.* Eurpean Commission. https://publications.jrc.ec.europa.eu/repository/handle/JRC128415?fbclid=IwAR2oU4AP-0aj8mp8sMfTLRgvC9ZpuaO7a942d-1b8UC3YCnC0bdnu88-G5XY

Wals, A. E. J. (2022). Transgressive Learning, Resistance Pedagogy and Disruptive Capacity Building as Levers for Sustainability. In GUNi (Ed.), *Higher Education in the World 8 – Special issue New Visions for Higher Education* (pp. 216–222). Global University Network for Innovation (GUNi).

WCED. (1987). *Our Common Future. World Commission on Environment and Development.* Oxford University Press.

Weldemariam, K., & Wals (2020). From Autonomous Child to a Child Entangled Within an Agentic World: Implications for Early Childhood Education for Sustainability. In S. Elliott, E. Ärlemalm- Hagsér & J. Davis (Eds.), *Researching Early Childhood Education for Sustainability: Challenging Assumptions and Orthodoxies* (pp. 13–24). Routledge.

Weldemariam, K., & Wals, A. (2020). From Autonomous Child to a Child Entangled Within an Agentic World: Implications for Early Childhood Education for Sustainability. In S. Elliott, E. Ärlemalm-Hagsér & J. Davis (Eds.), *Researching Early Childhood Education for Sustainability* (pp. 13–24). Routledge.

8 Facing the Future

Creating Learning Landscapes for Environment and Sustainability

A Changing Learning Landscape

A fundamental question this book has addressed is how the role of education has been changing over the decades since the 1970s. We have attempted to find some answers by tracking the development of narratives, thinking, and practice of learning and education in support of the environment and sustainability, through the decades. Along the way, we identified some clear differences in the way education and learning for the environment has been approached over time. In Chapter 2, we described an initial focus in the 1970s on *informing* young people (and later to adults) about the environment and on experiencing the wilderness with an instrumental focus on changing individual behaviour. This ran in parallel to efforts to *raise awareness* of and concern about global environmental problems. Moving into the 1980s in Chapter 3, there was a move towards clarifying and *understanding the science* of environmental issues with a strong technological focus. With time it became evident that increased understanding and awareness of these issues did not lead to the necessary changes or expected outcomes. Instead, the focus at the time was on problem-solving our way out of these concerns. Chapter 4 detailed the 1990s, where education evolved to a focus on more *interpretive and critical lines of inquiry* through increasing citizen engagement and participation. This pedagogical innovation extended in the 2000s, as learners became more actively involved in uncovering the root causes of social-environmental issues and exploring how these link to lifestyle choices. This tendency saw educators encouraging learners to *connect and change*, combatting the revival of instrumental interpretations of education that was more delivery-oriented and expert-driven. Chapter 6 highlights an increased focus on education systems and institutions during the 2010s that sought to *reframe and transform* our relationship with the environment. Chapter 7 on *regeneration and transition* gives attention to being aware of the structures and systems that influence how we think, feel, and act (see Figure 1.3 in Chapter 1 for an overview of these trends).

These are rough strokes in an impressionistic painting that represent a simplification of reality as the lines are often blurry, span different decades and

DOI: 10.4324/9781003467007-8

the colours mix. That said, the lines and colours we have drawn provide an opportunity for educators and policymakers to reflect on their own views and experiences of the development of education and learning over the years. It will also, hopefully, provide a platform from which to better understand the role of education and learning as we move into the future.

As the above trends highlight, our understanding and engagement with education and learning for the environment has changed substantially since the 1972 Stockholm Conference. What is clear is that the profile and presence of learning and education for the environment has been elevated in both policy discourses and communities of practice. The chapters have also tracked how it has become an established ambition in intergovernmental agendas on environment and sustainability as well as national education policies.

In this last chapter, we engage with an emerging *wildcard* in education – Artificial intelligence (AI). When we initiated this writing project in 2022, the wealth of information at times felt overwhelming. Going through databases and journals, reading articles and reports, has been a time-consuming endeavour, as was our attempt to identify, write, and structure knowledge and experiences into a trajectory of education and learning over decades. As we have noted earlier, this was a reflexive process involving discussion between the authors, as well as outside input. This has been a rewarding process, but also taxing in terms of weaving the perspectives, experiences, and writing styles between the authors.

Enter 2024, and an inescapable player has emerged in the field of education: Artificial Intelligence (AI). Launched in November 2022, the leading player is ChatGPT, which is a generative tool that allows users to enter prompts to receive AI-generated images, text, or videos that appear humanlike. As touted by many, AI has the potential to play a pivotal role in the fusion of digital and green agendas, through enabling smart environmental management, and driving green innovation for a greener future (Mondejar et al., 2021). However, there are concerns around data privacy breaches, the outsourcing of thinking to algorithms that often serve commercial interests, the exacerbation of inequalities, and environmental impacts of AI infrastructure (Khowaja et al., 2024). While early research has shown that ChatGPT has the potential to function as an assistant for educators, much attention has also been placed on the challenges in its use such as producing incorrect or false information and circumventing plagiarism detectors (Lo, 2023). Of course, there is another issue related to the use of IT and AI in everyday life and the world of work, including in academia: its enormous ecological footprint (Berthelot et al., 2024). Ideas are already being put forward for how to address these issues within educational practice and research (see Farrokhnia et al., 2024).

In line with one of the overarching themes of this book, we see in AI the continuation of the tension in education and learning for sustainability between a technological focus driven by efficiency, and an emancipatory focus that champions social equity and empowerment. While AI can optimise personalised learning and upscale educational efforts, it also has the potential to sideline critical thinking and perpetuate existing inequalities through widening the gap between those with and without digital literacy.

Recognising the importance of AI as an emerging technology, and simply curious as to how it works, one of the authors entered key emerging concepts like posthumanism, decolonisation, whole-school approaches (WSAs), intergenerational justice into the ChatGPT interface and asked the program to link them in a short 1000 word narrative about future directions for education in relation to the environment and sustainability. The text provided a remarkably comprehensive summary of the narrative we have been presenting in this book. There were obvious mistakes in the citations, and we the human authors felt at times that the AI-generated text was patronising and objectified learners. Ironically, the empowerment of learners is one of the core themes of this book and a core outcome sought by contemporary approaches to education and learning in the context of sustainability. We also noted that the AI-generated text lacked an appreciation of the complexity and the unexpected turns that occur in reality, perhaps something that is uniquely human. At the same time, we (and many of our peers) have also written texts in the past that perhaps could have benefitted from greater clarity, and sometimes simplifying matters can be helpful.

While the above points could be addressed through modifying the text, one of the authors felt deeply conflicted in allowing an algorithmic language model to write this section, stating that this felt like leaving the future to be written by technology; rather than the authors who needed to hone in their expertise to craft the chapter. After discussion between the authors, we collectively decided to work from the initial AI text as a foundational skeleton, with the logic that we would refine, deepen, and at times redirect the original AI text. In short, while ChatGPT initially helped us speed up the writing process for this chapter, it still required several iterations, modifications, and additions to make it work for the purpose of the book and its closing chapter. In the end, little of the original ChatGPT-initiated text prevailed. Instead, what remains is an original text of which ChatGPT got the ball rolling; a text that has been informed by grounded experience as well as academic expertise and that may be constrained by our own human blindfolds and cultural perspectives.

Future Directions and Characteristics of Education and Learning for the Environment

Education stands as the cornerstone of society, shaping the minds and hearts of future generations. Yet, as we navigate the complexities of the 21st century, it becomes increasingly evident that traditional paradigms of education are not only insufficient in addressing the multifaceted challenges that lie ahead but may also be directly contributing to divided and self-destructive societies (Shiva, 2013). The future of education not only demands a systemic approach –

one that encompasses the interconnectedness of individuals, communities, and the natural world – but also one that strengthens deep democracy, that invites critical thinking and diversity, as well as spaces for deliberation and transgression (Lotz-Sisitka et al., 2016).

The holistic perspective of nested systems continues to be an aspiration in learning for sustainability (Horlings, 2015; Ives et al., 2020). This perspective underscores the importance of nurturing not only cognitions and academic proficiency but also emotional intelligence, social consciousness, and ethical awareness. In the coming decades, this systemic view is likely to be extended as authorities help to carve learning landscapes that connect the learners in formal education with non-formal learning experiences found in their local and regional communities (Tilbury, 2024b). This approach brings the outside world into the school and higher education, as well as deepening the connections and capabilities of learners so they choose to engage in changes towards sustainability.

As the following sections will show, a transformative and holistic vision of education will need to prioritise equity and intergenerational justice as we resist societal paradigms that continue the trend towards social inequality and environmental degradation. We need to support and value educators who are teaching the next generation, promote pedagogies which resist the status quo through instilling hope, and provide students with the tools to effect change in their own lives and those of the communities they belong to.

A Living and Connected Sustainability

When asked to consider learning, many assume that this takes place in a classroom setting with the educator being the source of knowledge and leader of learning opportunities (Tilbury, 2024b). However, contemporary notions of education for sustainability displace the teacher from the centre of this activity bringing in non-formal educators into schools and seeking to connect the realities of the wider community and the everyday lives of pupils with academic learning. These views challenge traditional perceptions of education that have persisted over the decades, seeing it as a more dynamic and empowering process where other actors play important roles in shaping the knowledge, engagement, and abilities of learners.

In thesis sense, it is likely that buildings, playgrounds, and other components of the physical environment of schools will be increasingly seen as the focal point of learning as learning becomes a lived as well as a taught experience (Tilbury, 2024a). The idea of schools as places for contemplating, enacting, and recalibrating sustainability fits with the WSA that identifies the school, school grounds, and school building as a critical component of an integrated approach to sustainability (Mathie & Wals, 2022).

From a wider perspective, where learning continues to extend beyond classroom settings, we see learning for sustainability being increasingly prioritised in art, cultural and religious centres, as well as museums and municipalities, and businesses and NGOs. National and regional public bodies now have expectations that social learning opportunities will increase as new trainers, facilitators and educators join the quest for sustainable futures. Indeed, public bodies have started to incentivise the emergence of learning landscapes that support local or regional learning hubs for sustainability (Tilbury, 2024b). These landscapes connect non-formal education offerings to those available through formal, further and higher education so that there is alignment of learning outcomes but also quality control in the educational offerings.

The German city of Hamburg, for example, has established an ESD Masterplan for 2030, which offers a strategic frame from across the learning system involving museums, field centres, ecological markets, green festivals, the media as well as kindergartens, schools, vocational schools and institutes of higher education (BUKEA, 2022). The networking and interlinking of all areas of education is an elementary component of the plan – with over 120 different social actors and educators were involved in the development of the ESD Masterplan – as are inquiry or experiential-based learning approaches that root global issues in the local environment. This example shows how public authorities are increasingly recognising how non-formal offerings deepen the sustainability science and environmental expertise available to learners. At the same time, a focus is also placed on the need to provide a certification scheme to ensure the quality of educational and pedagogical underpinnings, as well as the building of green competences (Gonzalves & Tilbury, 2024). We anticipate the rise of learning landscapes for sustainability such as these across cities and regions that have prioritised sustainability as a social, economic, and/or environmental goal.

Similarly, society-oriented learning, as seen in the emergence of Living Labs in higher education (Tercanli & Jongbloed, 2022; Van der Wee-Bedeker et al., 2024), will continue to gain traction. This seeks to engage in co-creation processes between educators and multiple stakeholders to address social and organisational issues. While such collaborations can produce clear contributions to diverse knowledge, an important challenge is to explore how the competing values and demands from diverse stakeholders can be most effectively managed through the leadership of Living Labs (Tercanli & Jongbloed, 2022). Recent UNESCO dialogues under the Greening Education Partnership, for example, have shown how committed environmental advocates and specialists who are newcomers to education and learning have a tendency to reinvent wheels or revisit conceptions that have proven to be ineffective in advancing learning for environment or sustainability in the past.

Another channel that connects different stakeholders around real-life issues is citizen science, characterised by the active involvement of the public in scientific research. This has been discussed throughout this book, highlighting the importance of public engagement in environmental monitoring. From an early focus on the public solely gathering data, there is now more emphasis on citizen science projects to help drive environmental and social change through empowering individuals to exercise their environmental rights and responsibilities, with calls to promote intergenerational and intragenerational justice (Adamou et al., 2021).

Valuing the Role of Teachers and Teaching

The status of education, teachers, and the teaching profession directly correlates with the quality of education. It is paramount that teachers and teaching get the status and recognition they deserve; this will elevate the profession, attracting skilled individuals who are passionate about nurturing the next generation.

This issue is identified as a core concern of the Secretary General's High-Level Panel on the Teaching Profession that puts forward six core recommendations covering dignity, humanity, diversity, equity and inclusion, quality, innovation, and leadership and sustainability (UN, 2024). The report advocates for an enabling environment that promotes teachers to become catalysts of change in education and renew visions of education. The report acknowledges that climate change, biodiversity loss, unsustainable resource use, and persistent poverty exacerbate existing global inequities, threatening life on the planet. The report also recognises that climate change causes social and educational disruptions, disproportionately affecting those from disadvantaged and marginalised social groups (UN, 2024). Building upon the United Nations Transforming Education Summit (UN, 2022), The Global Report on Teachers (UNESCO 2024) has highlighted that for learners to think and act in transformative ways, as is required for effective learning in sustainability, teachers need to be adequately supported. This means decent working conditions, salaries and job security, reasonable workloads, formative assessments, and autonomy and agency. These aspects are all key to the delivery of quality education and learning for environment and sustainability and are thus critical to the future of the planet.

The Interplay between Climate Change, Gender, and Education

While we have seen the emergence of climate change education and the recognition by policymakers of the importance of education in climate change mitigation and adaptation, this text has not touched upon the way climate change can affect the education sector and its offering. Climate change poses a significant threat to education, manifesting in various ways that disrupt learning environments and exacerbate existing inequalities. One consequence of the growing frequency and severity of extreme weather events is the disruption of education systems worldwide. As hurricanes, droughts, floods, wildfires, and other disasters become more frequent and severe, schools are forced to close temporarily or even permanently, disrupting the learning process for millions of students (FCDO, 2023; UN, 2024).

As we can learn from research arising out of the COVID-19 pandemic, the impact of climate change-induced school closures will be particularly pronounced in vulnerable communities, where access to education is already limited (Grewenig et al., 2021). The children in low-income areas will be disproportionately affected, as they may lack access to alternative learning opportunities, such as online classes or private tutoring. This exacerbates existing educational inequalities, widening the gap between privileged and marginalised students (Agostino, 2010; FCDO, 2023).

Moreover, the disruption of education due to extreme weather events can have long-term consequences for individual students and entire communities. Extended school closures can lead to learning loss, hindering academic progress and future opportunities for affected students (Bayrakdar & Guveli, 2023). The displacement caused by pollution and climate-related disasters can uproot families, forcing children to change schools or drop out altogether, further disrupting their education and social development.

Redesigning cities, including school buildings and grounds, to be more climate-smart is a crucial step towards reducing the effects of climate change and creating sustainable environments. Incorporating green spaces into urban planning, such as parks, urban forests, and green roofs can help reduce the effects of urban heat islands, enhancing air quality, and offering recreational areas for residents (Elliott et al., 2020). The integration of green infrastructure into cities' planning and design can enhance resilience to climate change while promoting biodiversity and ecosystem services. At the same time, there is an increasing focus on engaging youth in climate action at the local and regional government level, such as the global network of Local Governments for Sustainability (ICLEI, 2023). Within a WSA, children and youth could become actively involved in the greening of school buildings and grounds as a part of their educational experience (Mathie & Wals, 2022). This would not only cultivate green skills and competencies but also facilitate learning across different generations (Tilbury, 2023).

Finally, incorporating climate justice entails confronting the unique obstacles faced by adolescent girls, especially in low- and middle-income countries and humanitarian settings. It is anticipated that more attention will be directed towards alleviating the marginalisation experienced by this demographic, given their heightened susceptibility to the disparate effects of climate change. The Gender and Adolescence Global Evidence (GAGE), a comprehensive study encompassing 20,000 adolescents in developing nations, has already documented the adverse ramifications of climate change on the educational, health, and socioeconomic prospects of young women (GAGE 2017). Communities residing on the frontlines of climate change are poised to gain prominence in educational discussions, as their capacity to acquire knowledge, prosper, and endure faces escalating challenges amid shifts in global climatic patterns (FCDO 2023).

Intergenerational Justice and Equity

Intergenerational equity is a concept that underscores the ethical dimensions of sustainability, highlighting the rights of future generations to the same natural resource base as the present generation. The UN SG's Declaration for Future Generations is expected to see the intergenerational equity position gaining strength. This recognises that the right to a healthy environment is on the line – a right owed to all people and non-human beings on the planet, including future generations – and extends the responsibility to those yet to be born (UN, 2023). Ethical and rights-based frameworks underpin this movement that seeks to connect our actions and inactions with long-term consequences that will affect those who will inherit the world (Davidson, 2023).

We will also see the 'good ancestor' principle (Krznaric, 2020) gain currency as well as the wave of youth-led climate litigation with support from around the world. However, there is an 'intellectual vacuum' with regards to what long-term thinking means tangibly for the education sector (Krznaric, 2020). Inspirations for filling this void could mean drawing upon the worldviews and practices of indigenous cultures, where the concept of equality among generations is deeply rooted. Native American knowledge systems are underpinned by the Seventh Generation principle that locates current decision-making and practices within longer-term frames. Another inspiration is international law, which recognises that all human beings, regardless of what time they live, are equal in dignity and rights. The Universal Declaration of Human Rights (United Nations, 1948) and the International Covenant on Civil and Political Rights (United Nations, 1967) are built on the notion that there is no temporal limit to these rights.

At the same time, a mounting challenge is how to effectively integrate the concern for future generations into education legislation systems and processes. As noted by Alemanno (2023), successfully incorporating the interests of future generations into contemporary policy involves more than just legal codification or creating new, often fragmented institutions and mechanisms. Instead, it necessitates a comprehensive, forward-thinking, and proactive approach by all governmental bodies. Inspirations for such approaches can be taken from the Dutch Manifesto for Future Generations (Generaties, 2024) and Welsh Protocol for Future Generations (2024), which are examples of advocacy and policy work that will make their presence felt in coming years and will likely influence education and learning for sustainability in the future. In terms of organisations, this can be seen by the focus on anticipatory governance, whereby the UN Summit for the Future is to address issues of intergenerational equity in global governance and establish a UN special envoy to advocate for future generations (United Nations, 2024). The UN Pact for the Future will enhance opportunities for younger generations to participate in shaping the future as well as seek ways of embedding the interests and rights of future generations in inter-governmental decision-making (UN 2024).

Although this concept in education is not new (see Tilbury et al., 2002), education is now expected to embrace the principle of intergenerational justice (Meijers, 2023; Tucker, 2008), acknowledging our responsibility to future generations. Embedding these concepts through curriculum and pedagogy, and at all educational levels from kindergarten (Oropilla & Ødegaard, 2021) up to students and the elderly (Gardner & Alegre, 2019) can contribute to learners appreciating the intricate web of interdependence between human societies and the natural world. While 'futures education' will gain further importance as a result of these developments, it will be pivotal to engage educators and students with debating concepts such as 'futures justice', and how these can be put into practice. In this respect, it is important to learn to not just anticipate but also craft alternative futures to the ones currently being faced.

Decolonising Education and 'Resistance Pedagogy'

The world has witnessed what systemic inequality has meant for those discriminated against because of their gender or race; the legacy has persisted through generations. It is imperative that education embraces 'the right' of intergenerational equity and long-term thinking in taking on the challenge to develop more inclusive and regenerative learning environments. This will involve promoting forms of education that seek fundamental change, advocating for transformative actions that challenge existing inequalities, and promoting social and environmental justice. One such form is decolonising education which seeks to dismantle oppressive structures, restore cultural integrity, and foster inclusive, equitable learning environments for all students. Moving beyond fairness, this encompasses broader societal concerns, including environmental sustainability and intergenerational equity. It further involves recognising and rectifying historical injustices, advocating for environmental stewardship, and promoting a culture of global citizenship and responsibility. In this way, we not only generate supportive environments where all individuals feel valued, respected, and empowered to fully participate in learning experiences, but also to challenge discrimination and marginalisation in all its forms.

Resistance pedagogies challenge and transgress traditional educational norms, advocating for social change and critical consciousness. These approaches reject passive learning in favour of active engagement with pressing social issues. They empower students to challenge dominant narratives, systems of oppression, and inequities and are particularly important to give voice to marginalised communities (Bajaj, 2015). Transgressive pedagogies push boundaries, disrupting conventional power dynamics and inviting marginalised voices to the forefront.

In the future, in addition to the shift towards alternative forms of teaching and learning and the redesign of learning environments (Anderson & Rivera-Vargas, 2020), conventional educational technologies, like curriculum materials in the form of school textbooks and educational resources will need to be recrafted to align with this critical perspective. In addition, there is the question of how resources for schools are generated and who is involved in the creation or generation process. Exploring such questions leads us to addressing the politicised agendas and power relations that seek to maintain the status quo in education. More participatory approaches to resource generation will be important so that the power relations that need to be challenged, also in the content of the resources, are also reflected in the way the resources are generated. This can help dismantle unsustainable processes in education.

The Posthuman and New Materialist Turn

Central to the paradigm shift in education embraced in this chapter is the realisation that humans are but one thread in the tapestry of life. Education must embrace more relational forms of learning that can help transcend anthropocentrism and help people realise or reaffirm the intrinsic value of all beings. Drawing inspiration from posthumanist perspectives and indigenous wisdom, educators can facilitate a deeper understanding of our interconnectedness with other species and the Earth (Braidotti, 2016; Lindgren & Öhman, 2019). Donna Haraway's 'Staying with the Trouble: Making Kin in the Chthulucene', provides a fruitful starting point for exploring these concepts (Haraway, 2016).

From an initial anthropocentric view of nature, one that has been in constant tension with ecocentric views already for many decades, we are currently seeing themes such as the rights of nature and posthumanism and inclusivity (interspecies equity) gain traction in educational debates (Lloro-Bidart, 2017). The coming years will likely see a recognition that humans are not the only species with exceptional qualities; rather, all species are endowed with remarkable attributes. The idea of decentering the human, developing some humility, acknowledging other species, and extending this acknowledgement also to matter, considering matter as vibrant and having agency as well, will attract more attention and yield corresponding forms of education. The posthuman turn also can be seen in attempts to give rights to nature (e.g. to rivers) and attempts to give voice to the human and the more than human, also in policymaking and decision-making processes (Hsiao, 2012).

Pedagogically speaking, we can find emerging niches like wild pedagogies (Blenkinsop et al., 2022) and Kimmerer's Braiding Sweetgrass (2013) as examples of immersive relational pedagogies that help decenter the human and can help learners to become more attuned to different ways of being and knowing. Often (outdoor) arts-based approaches play an important role. A key challenge will be to open up these possibilities not just to citizens who are in relatively comfortable situations and have the means to participate, but also to those living in more marginalised circumstances.

Societal Polarisation in an Age of Misinformation, Artificial Intelligence, and Fake News

Already in the 1972 Stockholm declaration, a strong reference is made in principle 19 concerning the dangers of misinformation to the environment:

> …It is also essential that mass media of communications avoid contributing to the deterioration of the environment, but, on the contrary, disseminate information of an educational nature on the need to protect and improve the environment in order to enable man to develop in every respect.

Now, 50 years later, the importance of mass media has only grown. We are currently witnessing immense political and societal polarisation around the world, in part fuelled by fake news and alternative facts spread through social media. It is therefore imperative that media take a more active role in separating facts from myths and distinguishing healthy doubt from intentionally cultivated doubt which is meant to confuse people or to delay action. The rise of AI exacerbates these issues, especially in education, as machine learning models can inadvertently perpetuate and amplify existing prejudices, potentially exacerbating inequality in educational outcomes.

From a technological perspective, robust countermeasures can be put in place through sophisticated fake news detection technologies (Choraś et al., 2021). As this book has argued, however, a more effective educational response is to enhance the agency of learners through promoting digital literacy as well as critical media literacy. This can help people understand the mechanisms behind misinformation and AI, encouraging citizens to engage in meaningful dialogues despite differing viewpoints. It is also likely that we will see 'quality' education for sustainability being more closely aligned to combating fake news and questioning of green consumerism claims. This could be accompanied by student interrogations of 'green' courses and learning opportunities, as educational expectations evolve (see as an example Ryan, 2024)

Reviving a Pedagogy of Hope

Last, education must empower students to confront the uncertainties of the future with courage. By integrating principles of active hope into the curriculum, inspired by the pioneering work of Joanna Macy and Chris Johnstone (2022), educators can instil a sense of agency and optimism in learners, enabling them to envision and enact positive change in the world.

At the same time, transformative change underpinned with agency, courage, and active hope needs to be accompanied with critical examination of power dynamics within education and within society in general. By fostering critical thinking and exploring decolonising pedagogies (Shahjahan et al., 2022), educators can challenge dominant narratives and create space for marginalised voices to be heard. Paulo Freire's 'Pedagogy of the Oppressed' (Freire, 1970) still provides a foundational framework for fostering emancipatory education that empowers learners to become agents of social transformation.

Cultivating hope also implies addressing people's so-called 'inner-sustainability'. This refers to the degree to which people have a state of comfort, well-being, and agency that makes them resilient and effective in coping with eco-anxiety in a generative way, if only to prevent numbing, apathy and simply withdrawing and giving up. Doing so will require the strengthening of people's transformative qualities which have only recently started to gain more recognition in the fields of sustainability science, education and psychology (Ives et al., 2023; Parodi & Tamm, 2018). Transformation entails shifts in people's mindsets and is fundamental to addressing many sustainability challenges, as was already pointed out by Donella Meadows who identified it as one of the key leverage points in realising a more sustainable world (Meadows, 1999) and has been reaffirmed more recently (Abson et al., 2017; Grossmann, 2019; Sterling, 2024).

When analysing some of the recent literature on inner sustainability and transformative qualities, certain attributes can be identified (see also Wamsler et al., 2018) as courage, agency and empowerment, compassion and relationality, openness, self-awareness and -reflection, and, finally, intrinsic values-based engagement. Paying attention to these through education, also in the context of (life-long) adult learning has implications for how we conceive of and design learning environments, including the physical design of schools and school grounds, as they will need to allow for ample safety, trust, inclusion and empathy in order for people to become comfortable with discomfort and anxiety on the one hand and to build confidence and capacity for making change on the other. There will also be implications for the professional development of educators as they will not only need to have those transformative qualities themselves, they need to become competent in strengthening them in the people, young and old, they are working with.

Conclusion

While the 1972 Stockholm conference was ground-breaking in recognising the interconnections between development, poverty, and the environment, subsequent years saw a reductionist paradigm emerge, based on formalised science and technology. The gradual emergence of the concept of sustainability demonstrated the interconnections and interdependencies inherent in the socio-economic and environmental challenges faced, and the need to promote systemic and transdisciplinary approaches to address them. As we look towards the future, we see the need for transformational, whole of society approaches that address shared environmental challenges. This will require urgent international action. Education and learning must therefore become more relevant and responsive to these planetary concerns.

There is a continued need to address global issues such as climate change, ocean acidification, loss of biodiversity, deforestation, and air pollution, as well as the need to employ all means and tools available to improve the health of our planet. Many of these issues were already flagged 50 years ago but the concerns have moved from the margins to the mainstreams of society and their urgency is now widely acknowledged. Education and learning are essential in helping citizens navigate their individual and collective efforts towards changing our systems and futures. In a sense, education is the ultimate form of mitigation as it plays a key role in building democratic eco-literate societies underpinned with a desire for socio-ecological justice and an ethic of care. In the years to come, however, education will need to be transformed to be able to take on this existential role at a rate faster than the pace of change we have seen over the last 50 years. Some might deem this too radical, but the current events affecting planet Earth are far more severe.

Such a transformation does imply transgressing some current trends that work against sustainability. Several of them have been highlighted in this book: the emphasis on testing and measurement, the cognition bias, the increasing gap between the poor and the rich, the continued anthropocentric way of thinking, the disciplinary silos that still determine much of what goes on in schools and universities, the erosion of deep democracy, the mantra of continuous economic growth, and there is more. But there are growing niches that are beginning to get traction in education and policy, as well as in the everyday lives of people fuelled by a sense of community, concern for the future, and the desire for meaning and a sense of responsibility. This applies not just for oneself but also for others – both present and in the future.

What certainly helps is that in the last 50 years, the profile and presence of education and learning for environment and sustainability has slowly been elevated in both policy discourses and communities of practice around the world. What is important now is that we do not take steps back, but instead continue to learn from experience and carve solid pathways for mainstreaming learning opportunities across all sectors, for all people, and as a key component of life-long learning. The challenge of sustainability will be ongoing: sustainability is not a destination where once arrived people can sit back and relax. It will require continuous experimenting, learning, reflecting, connecting, questioning, and recalibrating. Where will we be as a people, as a species 50 years from Stockholm Plus 50, when the year is 2072? The escalating sense of urgency will hopefully translate to an increased effort to improve education, learning, and capacity-building alongside a robust democracy, rather than a turn towards eco-totalitarianism or worse, totalitarianism stripped from any ecological consciousness. Through the creation and upscaling of living examples of education and learning creating inspiring communities and regenerative cultures that breathe sustainability in all aspects of life, and many of those already exist around the world, we can move away from dystopian futures towards more hopeful ones.

References

Abson, D. J., Fischer, J., Leventon, J., Newig, J., Schomerus, T., Vilsmaier, U., von Wehrden, H., Abernethy, P., Ives, C. D., Jager, N. W., & Lang, D. J. (2017). Leverage Points for Sustainability Transformation. *Ambio, 46*(1), 30–39.

Adamou, A., Georgiou, Y., Paraskeva-Hadjichambi, D., & Hadjichambis, A. C. (2021). Environmental Citizen Science Initiatives as a Springboard Towards the Education for Environmental Citizenship: A Systematic Literature Review of Empirical Research. *Sustainability: Science Practice and Policy, 13*(24), 13692.

Agostino, A. (2010). *Gender Equality, Climate Change and Education for Sustainability.* http://www.e4conference.org/wp-content/uploads/2010/02/Equals24.pdf

Alemanno, A. (2023). Protecting Future People's Future: How to Operationalise Present People's Unfulfilled Promises to Future Generations. *European Journal of Risk Regulation, 14*(4), 641–655.

Anderson, T., & Rivera-Vargas, P. (2020). A Critical Look at Educational Technology from a Distance Education Perspective. *Digital Education Review, 37*, 208–229.

Bajaj, M. (2015). "Pedagogies of Resistance" and Critical Peace Education Praxis. *Journal of Peace Education, 12*(2), 154–166.

Bayrakdar, S., & Guveli, A. (2023). Inequalities in Home Learning and Schools' Remote Teaching Provision During the COVID-19 School Closure in the UK. *Sociology, 57*(4), 767–788.

Berthelot, A., Caron, E., Jay, M., & Lefèvre, L. (2024). Estimating the Environmental Impact of Generative-AI Services Using an LCA-Based Methodology. *Procedia CIRP, 122*, 707–712.

Blenkinsop, S., Morse, M., & Jickling, B. (2022). Wild Pedagogies: Opportunities and Challenges for Practice. In M. Paulsen, J. Jagodzinski & S. M. Hawke (Eds.), *Pedagogy in the Anthropocene: Re-Wilding Education for a New Earth* (pp. 33–51). Springer International Publishing.

Braidotti, R. (2016). Posthuman Critical Theory. In D. Banerji & M. R. Paranjape (Eds.), *Critical Posthumanism and Planetary Futures* (pp. 13–32). Springer India.

BUKEA. (2022). *Education for Sustainable Development (ESD) Master Plan 2030.* Hamburg Ministry for the Environment, Climate, Energy and Agriculture (BUKEA).

Choraś, M., Demestichas, K., Giełczyk, A., Herrero, Á, Ksieniewicz, P., Remoundou, K., Urda, D., & Woźniak, M. (2021). Advanced Machine Learning Techniques for Fake News (online Disinformation) Detection: A Systematic Mapping Study. *Applied Soft Computing, 101*, 107050.

Davidson, J. (2023). *#FutureGen – Lessons from a Small Country?* Chelsea Green Publishing.

Elliott, H., Eon, C., & Breadsell, J. K. (2020). Improving City Vitality Through Urban Heat Reduction With Green Infrastructure and Design Solutions: A Systematic Literature Review. *Buildings, 10*(12), 219.

Farrokhnia, M., Banihashem, S. K., Noroozi, O., & Wals, A. (2024). A SWOT Analysis of ChatGPT: Implications for Educational Practice and Research. *Innovations in Education and Teaching International*, 61(3), 460–474.

FCDO (2023). *Addressing the Climate, Environment, and Biodiversity Crises in and through Girls' Education*. Foreign, Commonwealth & Development Office.

Freire, P. (1970). *Pedagogy of the Oppressed*. Continuum.

Gardner, P., & Alegre, R. (2019). "Just Like Us": Increasing Awareness, Prompting Action and Combating Ageism Through a Critical Intergenerational Service Learning Project. *Educational Gerontology, 45*(2), 146–158.

Generaties, L. T. (2024). *Manifest Toekomstige Generaties [Manifesto for Future Generations]*. Manifest. https://labtoekomstigegeneraties.nl/manifest/

Gonzalves, S., & Tilbury, D. (2024). *Non-Formal Education on Green Transition and Sustainability. EENEE Analytical Report 3/2024t*. European Commission.

Grewenig, E., Lergetporer, P., Werner, K., Woessmann, L., & Zierow, L. (2021). COVID-19 and Educational Inequality: How School Closures Affect Low- and High-Achieving Students. *European Economic Review, 140*, 103920.

Grossmann, I. A. A. D. (2019). Applying Wisdom to Contemporary World Problems. In R. J. Sternberg, H. C. Nusbaum & J. Glück (Eds.), *Wise Reasoning in an Uncertain World*. Springer Nature.

Haraway, D. J. (2016). *Staying with the Trouble: Making Kin in the Chthulucene*. Duke University Press.

Horlings, L. G. (2015). The Inner Dimension of Sustainability: Personal and Cultural Values. *Current Opinion in Environmental Sustainability, 14*, 163–169.

Hsiao, E. C. (2012). Whanganui River Agreement-Indigenous Rights and Rights of Nature. *Environmental Policy and Law, 24*(6), 371–375.

ICLEI (2023). *Unlocking the Power of Youth*. Local Governments for Sustainability. https://e-library.iclei.org/uploads/Unlocking_the_power_of_youth_Checklist.pdf

Ives, C. D., Freeth, R., & Fischer, J. (2020). Inside-out Sustainability: The Neglect of Inner Worlds. *Ambio, 49*, 208–217.

Ives, C. D., Schäpke, N., Woiwode, C., & Wamsler, C. (2023). Imagine Sustainability: Integrated Inner-Outer Transformation in Research, Education and Practice. *Sustainability Science, 18*(6), 2777–2786.

Khowaja, S. A., Khuwaja, P., Dev, K., Wang, W., & Nkenyereye, L. (2024). ChatGPT Needs SPADE (Sustainability, PrivAcy, Digital divide, and Ethics) Evaluation: A Review. *Cognitive Computation*. https://doi.org/10.1007/s12559-024-10285-1

Kimmerer, R. (2013). *Braiding Sweetgrass: Indigenous Wisdom, Scientific Knowledge and the Teachings of Plants*. Milkweed Editions.

Krznaric, R. (2020). *The Good Ancestor: A Radical Prescription for Long-Term Thinking* (Vol. 88, pp. 279–280). The Experiment.

Lindgren, N., & Öhman, J. (2019). A Posthuman Approach to Human-Animal Relationships: Advocating Critical Pluralism. *Environmental Education Research, 25*(8), 1200–1215.

Lloro-Bidart, T. (2017). A Feminist Posthumanist Political Ecology of Education for Theorizing Human-Animal relations/relationships. *Environmental Education Research, 23*(1), 111–130.

Lo, C. (2023). What Is the Impact of ChatGPT on Education? A Rapid Review of the Literature. *Education Sciences, 13*(4), 410.

Lotz-Sisitka, H., Ali, M. B., Mphepo, G., Chaves, M., Macintyre, T., Pesanayi, T., Wals, A. E. J., Mukute, M., Kronlid, D., Tran, D. T., Joon, D., & McGarry, D. (2016). Co-Designing Research on Transgressive Learning in Times of Climate Change. *Current Opinion in Environmental Sustainability, 20*, 50–55.

Macy, J., & Johnstone, C. (2022). *Active Hope (revised): How to Face the Mess We're in with Unexpected Resilience and Creative Power.* New World Library.

Mathie, R. G., & Wals, A. E. J. (2022). *Whole School Approaches to Sustainability: Exemplary Practices from around the World.* Education & Learning Sciences/ Wageningen University.

Meadows, D. (1999). *Leverage Points: Places to Intervene in a System.* Sustainability Institute.

Meijers, T. (2023). Climate Change and Intergenerational Justice. In G. Pellegrino & M. Di Paola (Eds.), *Handbook of the Philosophy of Climate Change* (pp. 623–645). Springer International Publishing.

Mondejar, M. E., Avtar, R., Diaz, H. L. B., Dubey, R. K., Esteban, J., Gómez-Morales, A., Hallam, B., Mbungu, N. T., Okolo, C. C., Prasad, K. A., She, Q., & Garcia-Segura, S. (2021). Digitalization to Achieve Sustainable Development Goals: Steps Towards a Smart Green Planet. *The Science of the Total Environment, 794*, 148539.

OCFGW. (2024). *Wales Protocol for Future Generations – From Declaration to Implementation.* Office of the Commissioner for Future Generations of Wales. https:// www.futuregenerations.wales/wp-content/uploads/2024/05/Wales-Protocol-for-Future-Generations.pdf

Oropilla, C. T., & Ødegaard, E. E. (2021). Strengthening the Call for Intentional Intergenerational Programmes Towards Sustainable Futures for Children and Families. *Sustainability: Science Practice and Policy, 13*(10), 5564.

Parodi, O., & Tamm, K. (2018). *Personal Sustainability: Exploring the Far Side of Sustainable Development.* Routledge.

Ryan, A. (2024) *Students Driving Curriculum Quality for Sustainability: A UK QAA Funded Initiative' Presentation at 19th UNECE Steering Committee in ESD*, Geneva: UNECE accessed 4th June 2024.

Shahjahan, R. A., Estera, A. L., Surla, K. L., & Edwards, K. T. (2022). "Decolonizing" Curriculum and Pedagogy: A Comparative Review Across Disciplines and Global Higher Education Contexts. *Review of Educational Research, 92*(1), 73–113.

Shiva, V. (2013). *Making Peace with the Earth.* Pluto Press.

Sterling, S. (2005). *Linking Thinking, Education and Learning: An introduction.* WWF-UK.

Sterling, S. (2024) *Learning and Sustainability in Dangerous Times.* London: Agenda Publishing

Tercanli, H., & Jongbloed, B. (2022). A Systematic Review of the Literature on Living Labs in Higher Education Institutions: Potentials and Constraints. *Sustainability: Science Practice and Policy, 14*(19), 12234.

Tilbury, D. (2023). *Sustainable Learning Environments in Schools: Rethinking Places and Spaces of Learning. Input Paper, DG EAC Working Group on Schools and learning for Sustainability.* Brussels: European Commission

Tilbury, D. (2024a). *The Twin Transitions: Digital and Sustainability Learning in Schools Working Group on Schools: Learning for Sustainability.* Brussels: European Commission.

Tilbury, D. (2024b) *Local Learning for Sustainability: Places, Partners and Participation Working Group on Schools: Learning for Sustainability. Input Paper, DG EAC Working Group on Schools and Learning for Sustainability.* Brussels: European Commission

Tilbury, D., Fien, J., & Stevenson, R. B. (Eds.). (2002). *Education and Sustainability: Responding to the Global Challenge.* IUCN.

Tucker, M. E. (2008). World Religions, the Earth Charter, and Sustainability. *Worldviews: Global Religions, Culture, and Ecology, 12*(2-3), 115–128.

UN. (2022). *United Nations Transforming Education Summit – Thematic Action Track 3: Teachers, Teaching and the Teaching Profession.* United Nations.

UN. (2023). *What is the Right to a Healthy Environment?* Office of the High Commissioner for Human Rights (OHCHR) United Nations Environment Programme (UNEP) United Nations Development Programme (UNDP).

UN. (2024). *UN Secretary-General's High-Level Panel on the Teaching Profession. Summary and Deliberations.* ILO UN UNESCO.

UNESCO. (2024) *Global Report on Teachers: Addressing Teacher Shortages and Transforming the Profession.* UNESCO. Global Report on Teachers: Addressing teacher shortages and transforming the profession | Teacher Task Force

United Nations. (1948). *Universal Declaration of Human Rights.* United Nations. https://www.un.org/en/about-us/universal-declaration-of-human-rights

United Nations. (1967). *International Covenant on Civil and Political Rights.* United Nations. https://treaties.un.org/doc/treaties/1976/03/19760323%2006-17%20am/ch_iv_04.pdf

United Nations (1972). Declaration of the United Nations Conference on the Human Environment. https://www.un.org/en/conferences/environment/stockholm1972

United Nations (2024, May 14). *UN (2024) Pact for the Future; First Revision May 2024.* Pact for the Future. https://www.un.org/sites/un2.un.org/files/sotf-pact-for-the-future-rev.1.pdf

van der Wee, M. L., Tassone, V. C., Wals, A. E., & Troxler, P. (2024). Characteristics and challenges of teaching and learning in sustainability-oriented Living Labs within higher education: a literature review. *International Journal of Sustainability in Higher Education, 25*(9), 255-277.

van Dijk-Wesselius, J. E., van den Berg, A. E., Maas, J., & Hovinga, D. (2019). Green Schoolyards as Outdoor Learning Environments: Barriers and Solutions as Experienced by Primary School Teachers. *Frontiers in Psychology, 10,* 2919.

Wamsler, C., Brossmann, J., Hendersson, H., Kristjansdottir, R., McDonald, C., & Scarampi, P. (2018). Mindfulness in Sustainability Science, Practice, and Teaching. *Sustainability Science, 13*(1), 143–162.